C000156560

Death machines

Manchester University Press

Death machines

The ethics of violent technologies

Elke Schwarz

Manchester University Press

Copyright © Elke Schwarz 2018

The right of Elke Schwarz to be identified as the author of this work has been asserted by her in accordance with the Copyright, Designs and Patents Act 1988.

Published by Manchester University Press
Altrincham Street, Manchester M1 7JA

www.manchesteruniversitypress.co.uk

British Library Cataloguing-in-Publication Data
A catalogue record for this book is available from the British Library

ISBN   978   1   5261   1482   2   hardback
ISBN   978   1   5261   1484   6   paperback

First published in hardback by Manchester University Press 2018
This edition published 2019

The publisher has no responsibility for the persistence or accuracy of URLs for any external or third-party internet websites referred to in this book, and does not guarantee that any content on such websites is, or will remain, accurate or appropriate.

Typeset
by Toppan Best-set Premedia Limited

This book is dedicated to Edith & Harry

Contents

Acknowledgements

This book is the result of several years' work and a great many people and places have helped shape it. The project began during my doctoral studies at the Department of International Relations at the London School of Economics and Political Science, where I was embedded in an intellectually rich and challenging environment. Many lively discussions, on campus and off, have influenced my thoughts on ethics, violence, and technology.

I am extremely fortunate to have had an exceptional supervisor whose ability as an intellectual guide, an outstanding scholar and a compassionate human being was instrumental to this book, so I want to begin by sincerely thanking Kimberly Hutchings for her unwavering guidance and mentorship. I am thankful also to Chris Brown, Patrick Hayden and Mervyn Frost, who have each provided great support and encouragement at different stages of the project.

For spirited debate, the honing of ideas and of course friendship, I am thankful to Mikey Bloomfield, Diego de Merich, Myriam Fotou, Kathryn Fisher, Joe Hoover, Marta Iñiguez de Heredia, Mark Kersten, Paul Kirby, Sebastian Lexer, Meera Sabaratnam, Laust Schouenborg and Nick Srnicek. I would also like to thank Andrew Futter and my other colleagues and friends at University of Leicester's HyPIR for providing such an encouraging and enabling environment in which to complete this work.

And then there are my colleagues, fellow travellers, and co-conspirators further afield. There is no way the book would be what it is today were it not for stimulating interlocutors at various workshops and conferences, many of whom inspired a thought or two that you might find in this book, specifically Wim Zwijnenburg, Peter Asaro, Heather Roff, John Emery and Caroline Holmqvist. I also would like to thank the editors at Manchester University Press, particularly John Banks for his meticulous copy-editing work, and the reviewers of the manuscript for their helpful comments.

I have also been fortunate enough to count on an astonishing support network closer to home. Specifically, I would like to thank all of the Sammans for their extraordinary generosity and magnificent writing retreat. My deepest gratitude is owed to my wonderful family for their unconditional support and inexhaustible optimism, and to Amin Samman, who, with his unfailing encouragement, his keen intellect and sharp eye, has been nothing short of spectacular.

Introduction: The conditioned human

Neither violence nor power is a natural phenomenon, that is, a manifestation of the life process; they belong to the political realm of human affairs whose essentially human quality is guaranteed by man's faculty of action, the ability to begin something new. And I think it can be shown that no other human ability has suffered to such an extent from the progress of the modern age.

Hannah Arendt, *On Violence* (1970: 82)

'What are we doing?' Hannah Arendt posed this question as a guiding theme for her reflections in *The Human Condition*. It's a simple, yet comprehensive question that carries both an ethical and a political demand. The question necessitates a pause for thought, a moment for reflection. It requests an evaluation of actions and the contexts within which they take place. This question is concerned as much with *what* is happening in the present as it is concerned with *why* this present might be as it is. In such a vein, this book is motivated by questions about the 'what' and the 'why' of contemporary technologies of violence and the underpinnings of their ethics. The emergence of new technologies for violent practices – from lethal drones to so-called 'killer robots', to weaponised Artificial Intelligence – presents a challenge to mainstream accounts of ethics in international political theory and raises important questions: what is this present in which technology stands poised to subsume humanity? And how can we recognise, decipher and understand more clearly what we are doing in this technological present? When acts of political violence become introduced as technologically justified practices, the need to interrogate the foundations that underwrite the politics and ethics of such violence arises anew. In 1962, Sheldon Wolin posed the question thus: 'Do the social and political forms of any given age constitute a particular method for adjusting to violence?' (2009: 39). He asked this question so that techniques to limit the unprecedented potential for violence in his time could adapt not merely to dealing with symptoms but with the causes.

This book is motivated by the perplexities of our contemporary wars, in which new practices and technologies of violence are presented as a more ethical and superior way of killing. This turn to argue for a more ethical way of killing in war is emphasised in debates on the use of lethal drones and the development of new autonomous weapons systems. The previous US administration under Barack Obama went to great lengths to characterise the use of lethal Unmanned Aerial Vehicles (UAVs) – more commonly known as drones – as ethical, lawful and prudent instruments in countering terrorism. Similarly, proponents of autonomous military robotics in the US Department of Defense (DoD) and elsewhere argue that the use of Autonomous Weapons Systems (AWS) – or killer robots – could make warfare in general more ethical and humane than in previous periods of human history. The emergence of ostensibly moral technologies of violence presents a challenge to mainstream conceptions of ethics in International Relations and International Political Theory. Current frameworks of just war traditions, ethics of war or international law, for example, all struggle to grasp, let alone challenge, the ethical implications of lethal drone strikes and the drive to establish killer robot armies. And where scholarly debates over the ethics of such weapons do take place, they are often confined to discussions of legality and effectiveness, ending up mired in problematic equations of fact with value. This impasse, along with the military discourse that surrounds lethal technologies, raises important questions about what is at work in the relationship between such technologies, their uses and the ethical justifications given for practices of political violence. In particular, what enables the framing of an instrument for surveillance and killing as an inherently ethical instrument? What kind of sociopolitical rationale underpins such a framing? And how does this rationale itself engender new regimes of high-tech killing? *Death Machines* addresses these questions by offering an analysis of how the production of techno-biopolitical subjectivities undergirds contemporary forms of technologised warfare.

In order to do this, I draw on the work of Hannah Arendt, who had an astute grasp of the biopolitical and scientific-technological implications of the modern human condition. To date, a range of scholars have drawn on Arendt for analyses of biopolitical dimensions of violence. However, a systematic account of her work on biopolitical trajectories and technologies remains underdeveloped in current scholarship. In the first part of this text, I establish such an account and I argue that the Arendtian analysis draws out a duality at work in the biopolitical shaping of subjectivities – the politicisation and technologisation of life itself on one hand, and the emergence of biological imageries that inform metaphors and processes of politics on the other. This helps us better understand how contemporary ethical frames of political violence are produced and shaped. The second part of the book is then concerned with the possibility of ethical thinking in a biopolitical present that is mediated heavily through technological interfaces and networks, specifically in modern warfare. My focus is on how modern subjectivities are produced through technological and biopolitical mandates, how such subjectivities shape contemporary understandings of politics and violence, and how these understandings, in turn, foster a type of ethics that supports increasingly technologised modes of political violence. In this way, my concern is to uncover how mechanisms of contemporary politics not only turn life and death into a technical matter but also impose limits on the way we conceive of and are able to contest the ethics of contemporary warfare. In short, by building an Arendtian biopolitics framework to situate a critique of contemporary conceptions of ethics of violence, this book offers two contributions: it supplements existing accounts of biopolitics as political rationales and offers a new way to theorise and disrupt justifications for technology-driven processes of violence in present-day warfare, such as the increased use of lethal drone strikes and the advent of AWS in war.

Postmodern perplexities

Motivated by the perplexities of a modern political life preoccupied with biological and reproductive processes, and largely under the sway of scientific-technological authority, in which the capacity to annihilate all life on earth had become a technological possibility, Arendt's chief aim was to understand humans in a specific sociopolitical context and their capacity and potential for political action therein (1998: 3). Her concern with understanding what it is we are indeed doing in the modern world comprised a range of perspectives and anxieties – some unique to her time, and others for which she proved to have her finger on the pulse of a future time, including matters of scientific and technological advancements that pose pressing challenges today. Arendt took part, for example, in a symposium on an emerging cyber sphere, held in New York in 1964. The symposium hosted a broad range of participants, from computer scientists to civil rights activists and, allegedly, 'at least one spy', with the aim of debating the 'cybercultural revolution' and its potential sociopolitical implications (Bassett 2013). Similarly presciently, in her text *On Violence*, published in 1970, Arendt considered the possibility of military robotics explicitly and commented on the potential political implications of autonomous robot soldiers in the not-too-distant future (1970: 10). Both aspects – the cyber sphere and autonomous and intelligent weapons systems – pose new and urgent political and ethical problems for us today.

Many of the puzzles relevant to Arendt's time have accelerated in our own time. Over ninety countries are in possession of military drone technology, many developing or acquiring lethal strike capacity for their unmanned vehicles; the pursuit of fusing human tissue with technological circuitry for an ever-more technologised and resilient 'Super Soldier' is long under way (Mehlman et al. 2013); the development of AWS with ever greater levels of decision autonomy is ostensibly inevitable (Ackerman 2015) and the development of human-level Artificial Intelligence (AI), and consequently artificial super-intelligence is quite possibly only decades away (Bostrom 2014). Individually, and combined, these technologies have the potential to transfigure both civilian and military life, customs and practices in dramatic ways. In particular, the emergence of technologies of violence that have normalised practices that hitherto had been considered immoral – such as the targeted killing of individuals as a prophylactic measure to combat terrorism – demands that we make it a priority to reconsider the ethics on which such practices rest 'from the vantage point of our newest experiences and our most recent fears' (Arendt 1998: 9). The speed with which new technologies for killing are developed and deployed in the war on terrorism makes this task all the more urgent today.

No longer officially termed 'the global war on terror', US operations relating to the fight against terrorism have been subsumed under the euphemistic moniker 'Overseas Contingency Operations' since 2011.[1] This represents a clear shift away from the emergency operation of war as a response to atrocities or in anticipation of an imminent attack, and suggests a much more enduring (military) administration for the control of risk, terror and contingencies. This is especially reflected in the use of drones for lethal strikes carried out by the CIA against targets in countries that are not officially engaged in war with the United States, including Yemen, Somalia and Pakistan. Contingency operations make a different approach possible. As John Kaag and Sarah Kreps have observed: 'Contingent targets emerge at unexpected moments in any variety of places. Targeting these individuals requires not mass invasion, but so-called surgical strikes, that are made without declaring war on a foreign state' (2012a: 280). Not large numbers of boots on the ground but rather a professional surgeon is called upon here. This type of medical incursion becomes the predominant mode of interventionist violence, which assists in forging new normalising narratives, in which assassination, as a practice, appears to have become a normalised foreign policy option. This has been an expanding practice since the Obama administration made drone strikes the interventionist tool of choice; since Donald Trump has become Commander in Chief, the use of lethal drones has expanded, and restrictions on their use have been loosened. All indications point toward a relatively unrestrained use of lethal drones by the US and its close allies in the near future.

The logic implicit in the fight against contingency in the war on terrorism presents an ethical struggle *per se*, wherein underlying and divergent value systems, narratives and administrative perspectives inform both the practices and the goals of warfare. At stake in this moralised battle is nothing less than humanity itself – not merely the survival of humanity but its values and, importantly, its progress. It is the threat to the corpus and advancement of humanity that both mandates new technologies designed for 'better', more ethical warfare and simultaneously unsettles established norms of what is morally permissible and impermissible. The use of lethal drones in the fight against terrorism is emblematic of the drive towards new forms of allegedly ethical warring. Posited as technology that can fulfil the tripartite liberal mandate to be 'legal, ethical and wise', as spokespersons for the Obama administration repeatedly declared (Brennan 2012a/b; Carney 2013), drones have since gained a dimension as weapons that no other new military technology has hitherto acquired – virtuousness in their own technological right, bestowed with the 'real promise of moral progress' (Statman 2015). At the time of writing, the US drone war regime has been expanding its reach for over a decade, producing ever more 'zones of war' (Walzer 2016: 14) and the debates remain heated as to whether lethal drones are the most ethical or an inherently unethical weapon of war. There is no simple answer to this, but, as Kaag and Kreps note: 'when it comes to war, if it is easy, it is probably not moral' (2012b).

Putting aside the question whether drones or other military technologies are ethical or not, my aim here is to look at whether the changing nature of military technologies makes us think about ethics differently; whether they widen or limit the scope for ethical concern and ethical deliberation about violence in warfare; whether they shift our thinking about violence as a political instrument. In short, the book is concerned with how techno-biopolitical subjectivities might shape our capacity to think ethically. In this, my first task is to ensure that our theoretical frameworks are robust enough to be able to understand our contemporary condition adequately. The second task is then to map the subject of concern – here the ethics of violence – on to these frameworks. Reaching across disciplines, this book adds to existing scholarship by first constructing an extended frame of biopolitics which considers the impact of technology on modern society, through the work of Arendt, with which then to identify and excavate the rationales that inform ethical considerations for new technologically driven practices of political violence, such as the use of drones and military robotics for lethal acts. The questions this text tackles are threefold: How do techno-biopolitical logics shape contemporary subjectivities? How is it possible that violent technologies are framed as inherently ethical? And what are the limits to ethical thinking in a technologically conditioned society?

In attempting to answer these questions, Arendt's diagnosis that '[m]en are conditioned beings because everything they come in contact with turns immediately into a condition of their existence' (1998: 9) is crucial. The human condition is thus a perpetually co-constitutive affair: ideas, structures, artefacts, rules, routines – all that comes into existence with the human condition – becomes part of the human condition and enters the world's reality. In turn, 'the impact of the world's reality upon human existence is felt and received as a conditioning force' (Arendt 1998: 9). This applies to biopolitical structures as much as it does to the technologisation of modernity. In their potential to shape human subjectivities, both have intermeshed consequences for what we understand our selves, our politics and our ethics to mean. In a highly technologically mediated society, this applies to everyday life as much as it does for matters of war. For the investigations in this book, I identify three co-constitutive and interlaced elements to the contemporary human condition: biopolitics, technology and ethics.

Biopolitical conditioning

In analyses of political violence, biopolitics refers to institutionalised mechanisms and discourses of power over the body and biological functions at the individual and the population level, whereby political government and life government are folded in with one another for the administration of life politics. In the master and meta-mandate to secure the health, prosperity, survival and progress of a population, biopolitics is inseparably entwined with concerns and practices of control, prediction, and prevention. It is also reliant on distinct technologies of security, which facilitate the norms and practices that come to govern societies. Contemporary analyses of biopolitics employ predominantly a Foucauldian perspective of the technologies of political power, in which government and life government become 'imbricated with one another' for the administration of life as politics (Lazzarato 2002). Where traditional sovereignty imposes its power on the general public, governmentality imposes a normalising generality on to the individual and society as a sociopolitical body. The biopolitical administrative technologies in Nazi Germany's totalitarianism represent the most radical example of such modalities, but contemporary forms of life management, such as biometric identification schemes for populations in conflict zones, like the Biometric Enrolment and Screening Device, or physiological screenings, at immigration controls also reflect this category. A new form of politics crystallises with the implementation of biopolitical modalities as the basis for governmentality: life and 'the political' conflate, war and politics merge; the mandate to secure the health, prosperity, survival and progress of a population becomes not only the master-mandate for politics but also its meta-mandate.

Drawing on Michel Foucault's analyses, the work of Michael Dillon has been instrumental in highlighting this relationship and its continued relevance in contemporary politics. In this context, Michael Dillon and Luis Lobo-Guerrero point out that a politics that claims the protection of life is simultaneously always also a politics that seeks to secure life – a politics of security (Dillon and Lobo-Guerrero 2008). In its aim to render life secure, biopolitics is inseparably entwined with concerns and practices of control, prediction and prevention and is reliant on distinct technologies of security that facilitate norms and practices, which come to (self-) govern societies. Such norms, practices and technologies range from surveillance policies and border control mechanisms to regulatory policies on dietary requirements, to such extreme punitive measures as extraordinary rendition, torture or – as of late – targeted killing programmes, for the 'security' of a population. Where the biopolitical logic leads to a demarcation between a population that is to be 'secured' and that which might pose a risk to a population's health, prosperity and the overall development of its internal processes, security technologies become the primary apparatus for the institutionalised aim to render secure what is fundamentally unsecurable: life itself (Evans 2013; Dillon and Reid 2009).

Investigating global governance in liberal modernity as global biopolitics, Dillon, together with Julian Reid, builds on the modern reversal of Clausewitz's observation, which deems war to be the continuation of politics by other means, and diagnoses biopolitics to be a continuation of war by other means, enabled by a myriad of technological inventions and institutions that liberal societies have come to accept and perpetuate as the norm (Dillon and Reid 2009). Developing this analysis of a biopolitical paradigm in contemporary modernity further, Reid's work analyses the biopolitical implications of the global war on terror in light of life as rendered both pacified and mobilised through various tactics and modalities of biopolitics. His assessment of liberal modernity characterises the twenty-first century human as one 'whose security is threatened by its refusal to question the veracity of its distinction between what does and what does not constitute a life worth living' (Reid 2006: 12). Dillon and Reid take their lead from Foucault when they recognise the radical indeterminacy of life, its underlying contingency, to be at the centre of what modern biopolitical modalities and *dispositifs* aim to control, if not eradicate, in an ever-present and never-ending contestation. And they interpret this continual contestation in biopolitics as a warlike struggle over the aporia of an inherent indeterminacy of life in a security-driven society. In such accounts of the biopolitical rationale, war thus becomes immanent to liberal society by two means: on one hand through institutional structures within liberal society that are informed by the originary military structures upon which technologies of disciplines and biopolitics were modelled in Foucault's analysis, and on the other through perpetual and pervasive power struggles over life's indeterminacies at various levels of society (Foucault 1991; 2004). It thus becomes part of the security apparatus to render life as technologically manageable as possible. Reid critically argues that modern biopolitical life is in essence a logistical life, 'under the duress of the command to be efficient, ... and crucially, to be able to extol these capacities as the values which one would willingly, if called upon, kill and die for' (Reid 2006: 13).

Where the efficiency and functionality mandate is paramount in a biopolitical rationale, the logic relies equally on the 'other' to the efficiency and functionality mandate – failure and vulnerability. As Brad Evans stresses in his study of liberal biopolitical terror, the political logics of biopolitics seek to ascertain predictable outcomes for an inherently unpredictable entity: life. The object of this logic is formed by 'precarious and vulnerable subjects' whose sheer biological conditions of mortality and finitude posit the central problem and concern of biopolitics (Evans 2013: 196). This, in turn, renders life a perpetually irresolvable problem. And precisely in this lies the conundrum, as Evans notes; the 'entire discourse on security is paradoxically underwritten by an appreciation that life can never be made fully secure' (Evans 2013: 196). The inherently aleatory, plural and contingent nature of humans, in co-existence with others, renders them at risk and stands in stark contrast to the desire to secure life. Where efficiency and functionality are requirements for the continued security of the life process of humanity, vulnerability and failure become dangerous imperfections that put life as such, as a political project, in peril. In the logic of always-immanent and contingent threats to human life, through aleatory and unsecurable elements, security strategies must first conceptualise and define the human as a biopolitical being for the management of contingency and risk avoidance (Evans 2013: 45).

This encompasses a precarious rationale: as biopolitics renders life problematic in terms of its potentialities, its inherently aleatory and unpredictable nature; in terms of its lack of certainty, its vulnerability in finiteness and mortality, it is not only rendered perpetually at risk but also poses a continual risk (Evans 2013: 87–90). Evans frames this perpetual risk in terms of terror. This terror contained within life, the terror of the unpredictable, is thus woven into the very fabric of biopolitical life as a ubiquitous threat. It is, according to Evans, a latent terror that is contained in the tension between the securitisation mandate that seeks to ascertain life, and the inherent unpredictability and volatility of one's existence in the world (2013: 30). However, the perception of unpredictability, uncertainty and vulnerability as a perpetual threat is in itself conditioned by narratives which stipulate that certainty, security and control over aleatory processes can indeed be brought about, and only then is every potentiality perceived as lack of control, becomes a threat, and turns into latent terror. And as Evans notes, in a liberal political context we seek to mitigate this terror with violence as a political strategy, as a 'creative' solution to eliminating and reducing threats through technological prophylaxis, whereby drones and other automated and autonomous military robotics serve as a panacea for all such problems by enabling their violent eradication.

The biopolitics–technology complex that provides the technological ecology within which biopolitical subjectivities are shaped is crucial here. The human in a technology-driven biopolitical age is not only determined by rationality but first and foremost captured in scientific terms and rendered analysable, predictable and knowable. In his writings and lectures, Foucault engaged predominantly with technology as *dispositifs*, as institutions and mechanisms of power, and was interested to a much lesser degree in the material aspects of science and technology as biopolitically informed and working upon the world. Some have critiqued the Foucauldian concept of biopower as relying on a thoroughly outmoded understanding of how technology – material technology – functions (Braidotti 2011: 329; 2013: 117; Haraway 1997). This is reflected in many contemporary engagements with biopolitics. While literature drawing on Foucault's *dispositifs* for the (self)-control and management of populations addresses the techniques relevant for, and used in, securitisation practices in the context of war, it engages little with the very material aspects of rapidly developing technologies and their permeation of the sociopolitical (Western) realm. As Rosie Braidotti points out, there is, indeed, a discrepancy between Foucault's biopower and the contemporary structure of scientific thought (Braidotti 2011: 329). The contemporary structure of scientific thinking is significant in the biopolitical context, as it conditions the biopolitical human subjectivity. To date, scholarship that looks at the biopolitics–technology–violence nexus in terms of both biopolitically *and* technologically constituted subjectivities, and the ethical justifications they produce for violence, is rare. Especially accounts interrogating biopolitics and its relation to material technologies have remained sparse.[2] I argue that Arendt has usefully engaged with the structure of scientific and technological thought in a life-politics-centric modernity and her thoughts offer a way to analytically access the co-constitutive nature of biopolitics and material technology for the examination of the ethics of political violence today. Through her work we can better understand the techno-logical conditioning of the biopolitical subject and the acceptance of specific modes of political violence.

Technological conditioning

Recognising the immense potential of the impact of technology, Arendt neither condemned nor condoned scientific and technological developments as such, but was critically concerned with the political question of the use of these technologies. 'What we are doing' with the capacities of new technologies and scientific advancements, set within a biopolitcal context, not only is a 'political question of the first order' but, when it comes to warfare and practices of political violence, becomes also a pressing ethical question (Arendt 1998: 3). Arendt's broad yet detailed inquiry into both biopolitics and the perils of a technocratic society renders her a rich resource for the continued 'project of understanding' of our biopolitically informed modernity (Parekh 2008: 6). Arendt presciently, and perhaps speculatively, engaged with questions about technology and technology's impact on the human in various lectures and essays during the 1950s and 1960s.[3] Today it is clear that technology has a considerable impact on human life, politics and warfare and there is a growing body of work seeking to investigate the influence of technology on these spheres of life. Scholarship that looks specifically at the influence and impact of ever-accelerating technologies on contemporary politics and society has only begun to blossom within the last ten years.

From communications technologies to the implantation of microchips into brains to improve performance and brain activity to the use of remote-controlled, unmanned weapons systems to super-intelligent AI, technology is advancing at a pace that exceeds the political, legal and ethical frameworks upon which we have hitherto built our co-existence in a shared world. While the interplay of humans and machines has a long history, there has been a change in the hierarchical relationship that ensues. Humans no longer merely create their machines, but are increasingly constituted by them, as humans and machines merge faster than ever (Coker 2013: xv). As Christopher Coker observes, in the context of new technologies of warfare, at stake is no longer the 'interface of the human being and technology' but rather 'the integration of technology into the human being. This is something that is new' (2013: xv). The mutual integration of human and machine is new because it places the human in charge of the technological progress not only at the periphery of the human but of humans themselves, as if 'man had been suddenly appointed managing director of the biggest business of all, the business of evolution – appointed without being asked if he wanted it, and without proper warning and preparation' (Agar 2010: 3). It is new because it shapes us as humans, and, in turn, shapes our human interactions. It is also new because it requires us to urgently rethink what it means to be human in an anthropo-technical context in which, as Peter Sloterdijk suggests, 'technology puts humanity at risk but will also save humanity by creating superior human beings' (Sloterdijk 2009). Indeed, the question arises whether, in this anthropo-technical context, we are faced with a different, a new configuration of biopolitics and the violence this facilitates. An Arendtian perspective of the technologically informed logos of biology – bio-logos – not only as the basis for human subjectivity but also as the basis for conceptions of political practices provides us with the tools to understand the biopolitical rationales at work in this.

Inscribed in the technological logos is a scientifically informed biological logos of processes, which Arendt picked up on when she notes, referring to Werner Heisenberg's 1958 essay *The Physicist's Conception of Nature*, that technology appears as a natural ecology, a biological process, rather than a human artefact toward greater power (Arendt 2006b: 274). It is the technological mindset that produces ideas of 'nature as an expression of a universal machine, an algorithmic immanence', and simultaneously provides the blueprint for ever greater simulations of natural, biological processes (O'Connell 2017: 76). It is in this assimilation of living entities with technological processes that the merger of the two can advance and 'allow us to transcend [the] limitations of our biological bodies and brains' (Kurzweil 2016: 148). This is the foundation of contemporary forays into military robotics.

In other words, the political economy of technologies of violence and biology is entwined. The interplay of biology and technology suggests that political subjectivities are shaped along corresponding lines. We conceive of computers, machines and technology as logical extensions of (limited) human sensory and physical capacities in an ever-wider realm of applications, including warfare. The all-pervasiveness of the technological environment, modelled on and within the human logic, is now shaping and conditioning the human logic in return. Not only are human subjects and their subjectivity framed in bio-technological terms of understanding biological processes as computational processes. In this view, we are information systems, in both body and brain functions, whereby the depersonalised and deindividualised logos of machines comes to be seen as an ideal. But also, collectives of people are conceived of in abstracted technological terms and within a symbolic order of codifications and mathematical signifiers as repositories of information and codes in a cybernetic assemblage. This, as Evans points out, has 'profound bearings on the question of what it means to be a living thing, as life is seen to be able to generate beyond itself' (2013: 72). Braidotti makes a similar point and notes that 'the zoe-centred egalitarianism that is potentially conveyed by the current technological transformations has dire consequences for the humanistic vision of the subject' (Braidotti 2013: 141). In short, we should train our analytical lens on how we understand ourselves as humans and among other humans in the technological production of life and the contemporary merger of the human with machines.

In his critique of techno-subjectivities, Jean Baudrillard notes that 'man and machine have become isomorphic and indifferent to each other: neither is other to the other' – machines that, in Baudrillard's assessment, promote homogeneity, reproduction, replacement, prevention and prophylaxis for the 'technological purification of bodies' (2009: 143, 68). Not only is technology biologically informed but, in this, it also poses the ideal and politically powerful means to enable the logos of prevention and prophylaxis so crucial to the securitisation mandate of a biopolitical modernity. A growing technological drive thus also promotes an ever-greater drive towards preventive and prophylactic practices. Yet biopolitical technology relies on a logic that exceeds humans and their capacities. And it is here that a peculiar turn in contemporary modernity appears to take place, as the hierarchies of the human and machines shift and the two begin to fuse not merely performatively but also functionally and philosophically (Coker 2013: 18). This strand of thought is picked up again in Chapter 6, where I suggest that new technologies are increasingly placing the human in a space of functional obsolescence, or, rather, humans place themselves in such a space, specifically in the context of war and political violence. Technology, shaped and informed by scientifically established biological and neurological patterns, is hailed to be able to perform human functions more accurately, faster, more efficiently and, as some commentators posit, also more ethically. In this rationale, humans are grasped not only in their essence as biologically ascertainable beings but also as fallible in their aleatory humanness, and, likewise, humanity as fallible in its inherent alterity and unpredictability. The rationale of technology seeks to mitigate these fallibilities and flaws to ensure the continued performativity, functionality, progress and process of humanity and humans therein, tacitly attesting to an notably diminishing belief in human judgement and human choice in the aim to secure life (Coker 2013: xvii). Encapsulated in this is, however, a diminished horizon for meaningful political action – that is to say political action not as management and administration of populations and resources but as acts and practices of interaction through 'communication between singular entities and collective assemblages' (Braidotti 2011: 341). This curtailment of political action is made manifest through set of limits to plurality (as technology relies on homogenisation), a limit to language (as technology relies on the 'language' of code and abstraction) and a limit to contestability (as technology is posited not only as politically and ethically neutral but superior) in the quest to make the human and humanity more efficient, functional and secure.

The accelerated use of and dedication to the employment of lethal drones exemplifies this prevention mandate, and the underlying rationale for the use of drones reflects the biopolitically informed technology drive, as technological fixes become ever more attractive in the securitisation mandate. The dominant mindset in Washington and Pentagon circles clearly reveals such thinking. I discuss this more thoroughly in Chapter 7. Firmly held by proponents of the technology to be 'ethical and effective',[4] drones are framed not only as technologically superior to traditional weapons systems, in terms of efficiency and effectiveness for the achievement of goals in warfare, but also as performing acts of war more morally, valuably and wisely. Drones are, so the argument stipulates, able to conduct necessary acts of violence with better precision, greater ability to discriminate in terms of targets and with less human cost (in terms of both lives and money). They offer the ideal technology to take out threats and combat human evil before its risk-infused potentialities become realised. They are, in short, the ideal tool to technologically secure the very processes of a biopolitically conditioned and technologically informed sociopolitical body. Extending this logic further, proponents of lethal AWS suggest that such machines could conduct the act of necessary killing more efficiently and – already alluded to – more ethically.

The techno-biopolitical subjectivity and ecology provide the cartography for a securitisation rationale that is difficult to challenge or contest, ethically and politically. Not only does such human-centric technology condition human subjectivities toward a greater acceptance of a technology outlook or mentality, in creating a greater ontological reliance on analysability, predictability and the production of certain outcomes, but it also conditions the technological ecology within which acts of political violence are framed as necessary technical acts for the securitisation of progress and survival for the human and humanity at large. Evans argues that the imposition of 'moral imperatives on a society so that certain productive ways of living become normalised to the point that they are not even questioned is a sure way of embedding secure practices. Such moralisation is what allows biopolitical practices to take hold' (2013: 198). Violent biopolitical practices of securitisation, morally mandated by the need for survival and the continuation of progress for humanity, facilitated by techno-thanatological weaponry that is rendered as inherently ethical, should give us pause to think about justifying narratives of these practices. This, in turn, should give us pause to think how the biopolitical-technological nexus that shapes human subjectivities in technological terms informs certain contemporary perspectives of the ethics of political violence.

Ethical conditioning

In her work, Arendt sought to engage with the problem of evil and the unmooring of morality in a secular modernity – specifically against the background of the murderous regimes in Nazi Germany (Canovan 1995: 156). However, her work is largely void of distinct ethical theories, even though there is a tendency among her readers to consider her as a 'moral *thinker*' (Kohn 1990: 105). The absence of ethical theories in her work does not, however, mean that she was unconcerned with issues of ethics and morality.[5] On the contrary; throughout her work, she thought through the problems of politics, revolutions, totalitarianism and the human condition as such, continually concerned with the relationship between morality, law and politics in the private and public spheres. Having witnessed the 'total collapse of the "moral" order not once, but twice', Arendt grappled with the fact that people in Nazi Germany could adjust to a different set of moral principles upon which it seemed perfectly acceptable to do what had hitherto been forbidden and illegal, and, once the tables had turned, and the Nazi regime had been defeated, that ideas of moral right and wrong would change once again without much effort or any lengthy indoctrination (Arendt 2003a: 54). In other words, the ontology of morality is tested when that which was hitherto a crime has become socially sanctioned as a legal act (Bauman 2012: 210). This extreme form of adjusting from normality to abnormality and back was a clear indication for Arendt that neither personal nor political morality was sufficiently rooted and neither religion nor philosophy could avert the dangers of 'moral nihilism' (Canovan 1995: 174). In this she identified a crucially modern turn wherein morality is reliant on, and always temporarily enshrined in, rules, customs, regulations, codes and guidelines. Only in her later life, during her work on *Life of the Mind*, did she begin to more deeply address the core of the problem of morality, a task she never managed to complete before her death in 1975. Although her discussion on morality and politics remains incomplete and, perhaps for that reason, strongly contested (Butler 2009a; Kateb 1984), her considerations point to a precarious relationship between ethics and politics in modernity, which can be seen in the legitimising of practices of acts of political violence, such as targeted killings, previously considered to be immoral and illegal (Butler 2009a; Kateb 1984).

Even though Arendt's work at this juncture cannot offer theoretical analyses that speak directly to the specific problems of the ethics of political violence today, the biopolitical framework built from her work serves to examine the biopolitical rationales that inform such modern forms of ethics and the human subjectivities upon which they rest. The inexorable relationship between human subjectivity and ethics requires that we make humans and how human are constituted, and constitute their environment, a distinct part of the analysis of ethics (Campbell and Shapiro 1999: xi). Specifically where the continued merger of the human with technological innovations produces a biological-technological subjectivity that shapes the human condition, and humans are conditioned in turn, the question arises how this further conditions our engagements with others in terms of ethics, as well as what types of ethics and politics this produces. As scientists strive to 'investigate ways in which the human and machine may co-evolve, both functionally and performatively, and how we may even be able to biologically re-engineer ourselves' (Coker 2013: xiii), the effects this biopolitically informed techno-subjectivity might have on the ethics of political violence are paramount. And if we accept that 'modern ethics is a species of the metaphysics of subjectivity' (Campbell and Shapiro 1999: xi), and that biopolitically conceived and shaped technologies condition human subjectivities today, it becomes clear that ethics, as conceived in modernity, must be examined against the biopolitical background of the techno-subjectivities that condition the human in modernity.

Ethics in international politics has to date predominantly been conceived in terms of applied ethics (Nardin 2008: 15) and is chiefly concerned with the search for 'a singular ethical theory that could be devised in the abstract and applied in the concrete' (Campbell and Shapiro 1999: viii). While a number of post-structuralist scholars sought to address ethics in terms that consider aspects of contingency, alterity and potentiality, particularly so in the late 1990s, the events unfolding in the aftermath of the terrorist attacks on 11 September 2001 appear to have initiated a drive toward thinking of ethics in international politics in more practical terms, giving priority to the application of ethical principles in and of warring. Such practical approaches often mirror scientific processes, or algorithmic logics, in trying to secure 'correct' outcomes. This is represented particularly in the hyper-rational logic of current discourses on the ethics of war. Consider, for example, how Yitzhak Benbaji's analysis of self-defence as a convention of war engages with the subject through an algebraic syntax, setting the parameters for his computational analysis as follows: '[s]uppose that at a time t+1, Y poses a threat to X's life, aiming to eliminate an unjust threat posed by X at t' (Benbaji 2008: 477). Benbaji then continues to posit various IF/THEN propositions by which certain laws of wrongful or justified actions are ascertained.

The algorithmic considerations of acts with ethical content rely on a law-like structure in ascertaining ethical conduct. Here, this is the laws of mathematical reasoning. Enshrined in codes, laws and regulations, correct ethical conduct can thus be sought through algorithmic reasoning and abstracted for the application in a range of national and international political contexts. Discussion of the ethics of war, humanitarian interventions and most recently the use of automated military technology reflects such an ethico-legal framework for ascertaining which acts are ethical. This, again, is also a pertinent aspect in discussions on the use of lethal military robots that are capable of increased levels of decision-making functions, including the decision to target and kill humans. Advocates and producers of lethal autonomous robots suggest that installing an 'ethics module', which would faithfully incorporate the laws of war and other relevant international legal frameworks, would render war more humane and less messy (Arkin 2009b, 2010). According to this logic, robotic kill decisions could be made more safely and humanely and therefore, ethically. The logic of an ethics module is reliant on a conception of ethics as codifiable, as ascertainable, and as producing clear, secure and, ideally, certain outcomes. In our biopolitical and technological context today then, not only do we seek to secure humans, in their aleatory, vulnerable and uncertain characteristics, but ethical relations between such humans must also be ascertained, secured, made controllable. Such ethics is both limited and limiting. It is ethics as a mere technical problem.

The reliance on codes and the structures of law in the ethico-legal construct of this practical variant of ethics has the potential to eclipse an engagement with that which cannot be secured in ethics, with the actual content of the moment of ethical interaction and an engagement with concerns that do not fit into the programmes of ethical administration conceived in any ethico-legal constructs. Where the ethical language is simultaneously also the language of code, subordinated to the attainment of a specific ethical end, it becomes altogether meaningless, or paralysed, in Agamben's terms (2000: 116). As Coker notes, '[l]anguage matters, otherwise one becomes like Hardy's Jude who thought that you could understand Greek if you cracked a simple code in the professor's safe keeping' (2013: 297). Language matters also, because it is the means with which political contestation can ensue, and with it, ethical contestation. Where language is replaced by the 'meaningless formalism of mathematical signs' (Arendt 2006b: 274) we lack the proper political and ethical language to contest the contents of ethics and its meta-ethical underpinnings meaningfully, for 'speech is what makes man a political being' (Arendt 1998: 3). Here again the tension between politics as management and politics proper is relevant. Where politics is understood as action, ethics is implicit. For Arendt, politics arises from the encounter with the other, in a public context. In this encounter lies not only the potential for politics but, within it, the demand for ethics also arises. Melissa Orlie perceptively traces this relationship in her critique of contemporary conceptions of ethics:

> The determination of ethical conduct ... emerges, at least in part, through thinking with others not as a thought experiment but as a locatable political practice. Ethical conduct requires more than thinking about the limits of the self, it also demands ethical political work on those limits. (Orlie 1997: 196–7)

Applied ethics and an ethics that resides in the unmediated plurality of the encounter seem to be located at opposite ends of ethical sensibilities and practices, and the techno-subjectivity shaped by biopolitical rationales gives precedence to the former rather than the latter in seeking to secure ethical outcome. An ethics of encounter must take into consideration the very unpredictability that arises out of political and social situations, each in its own context. It is thus that Campbell and Shapiro suggest, along with others, that international political theory 'should promote an ethics of encounter without commitment to resolution or closure', whereby 'our responsibility to the other' should serve as the basis for ethical reflection (1999: xvii). Judith Butler frames it similarly when she suggests that '[e]thics is less a calculation than something that follows from being addressed and addressable in sustainable ways' (2009b: 181).

This book aims to open a space to rethink the ethics of certain practices of war and counter-terrorism today by recognising, examining and critiquing the biopolitical perspective that facilitates a human techno-subjectivity which views ethics either in code or as technologically neutralised, which may allow us to move beyond the biopolitical subjectivity condition. There is much to be done in rethinking ethics in contemporary international politics in order to avoid the paradoxical condition that ethical considerations are relegated to the margins of moralising international politics. To conceive of tangible solutions must remain a future aspiration and motivation for continued research and engagement with the topic. It is far beyond the scope of this book to offer concrete suggestions as to what the ways forward should be. This is for another time. What *Death Machines* seeks to accomplish, however, is to provide a pressing critical perspective as a point of departure for further inquiry.

Structure of the book

The book's chief concern is with how techno-biopolitical rationales condition us as contemporary subjects, on one hand, and to examine and critique the emergence and meaning of the ethics of violent technologies, on the other. It is through this type of interrogation that 'we can set aside the universalizing moral entrapments of liberal humanism, which reduces political ethics to a question of relations among already compliant political subjects' and aim to think through what it is that we are, in fact, doing (Evans 2013: 11). In this spirit, the book seeks to examine how the biopolitical rationale, set forth through an Arendtian account, informs politics, violence as politics, ethics and the shaping of human subjectivity. The book unfolds in two parts and eight chapters. The first part, constituted by Chapters 1 to 4, engages in an Arendtian exposition to posit the theoretical framework for the book's central analysis. The second part then departs from an exposition of Arendt's work and draws on the biopolitical framework established in order to examine how the techno-biopolitical rationales shape the ethics of certain, technology-driven, political practices today.

Chapter 1 places Hannah Arendt as a biopolitical thinker and establishes an Arendtian framework that unfolds the biopolitical analysis contained in *The Origins of Totalitarianism, On Revolution* and, importantly, *The Human Condition*. The chapter establishes key terms and concepts and extracts two crucial trajectories relating to life politics in Arendt's work: the prioritisation of life processes in modern politics and the mathematisation of life and its processes to yield biologically analysable and calculable subjects and subjectivities. In this, the chapter sets out to establish the crucial tension that emerges in modernity when the cyclicality of life processes, upon which biopolitical practices are modelled, meets with a mandate for the progress of the human and humanity. Finally, I address some of the objections and scepticisms of Arendt as a biopolitical thinker in this chapter, and stress that the conceptual understanding of 'politics' in Arendt's work must be distinguished from a contemporary and already biopolitically informed apprehension of the term as governmental management and administration.

In contemporary scholarship the notion of biopolitics is typically associated with the late work of Michel Foucault. An exploration of Arendt's analysis of modern biopolitics serves less as a corrective than an additional dimension to offer tools with which to understand the shaping of contemporary, technological, subjectivities. Chapter 2 therefore delineates where Arendt and Foucault intersect and where they diverge in their analysis of biopolitcs. Here, I first map out parallels in their engagement with the shift of life into politics, linking these to their respective use of other key concepts (such as essence, telos and power). This then serves to establish where Arendt is useful for additional insights that exceed Foucacult's work and highlights the relevance of Arendt's threefold insights to this discussion, stressing how such a view provides for a form of biopolitical analysis that concerns more than mere governmental management or administration.

With a conceptualisation of the shift of life into politics in Arendt's work set up, modalities of both politics and violence can then be situated and understood more clearly within the context of a biopolitical modernity. Chapters 3 and 4 establish the consequences of the techno-biopolitical rationale for the possibility for politics, and the role that violence plays in a biopolitical way of understanding politics. In both chapters, Arendt continues to be an instrumental interlocutor to stress the tension that emerges in a biopolitically informed political modernity that understands politics as professional administration, contrary to an Arendtian understanding of politics proper. Chapter 3 unpicks the anti-political essence of biopolitics by engaging with key aspects that are necessary for politics in the Arendtian account and that become inhibited and obstructed in a biopolitical modernity for the possibility of politics. The main argument I make in this chapter is that, in such a regime, the possibility for political action and contestation is narrowed through mandates that oppose key aspects of politics, such as plurality, speech and a tolerance of uncertainty. I also highlight how such impediments to politics proper promote the emergence of political violence as a means of producing predictable outcomes. Chapter 4 extends the argument that the anti-political essence of contemporary biopolitics opens up a pathway for instrumental violence, enabling violent practices to appear as expedient tools in the administration of humanity. In this chapter, I focus more closely on the place of violence within techno-biopolitical regimes. I begin by excavating the linkages between Arendt's biopolitical analysis and her writings on power and violence. I then argue that the anti-political essence of contemporary biopolitics opens up a pathway for instrumental violence, enabling violent practices to appear as expedient tools in the administration of humanity. Finally, I suggest that such an understanding of violence-as-politics overlooks the futility of violence as a political practice in important but rarely acknowledged ways.

The violence-as-politics instrumentality reverberates in ethical justifications of violence. It is here that it is necessary to transition away from

Arendt to explore the contemporary status and ethos of the ethics of political violence. In Chapter 5, I engage in greater detail with the specific form of ethics that is produced in a biopolitically informed sociopolitical context. The chapter delineates how aspects of the biopolitical framework permeate and shape justifications of violent acts and the ethics of war and armed conflict. It further problematises the conception of ethics as predominantly conceived in practical terms, as a form of applied ethics, framed in terms of code and algorithmically ascertainable rules of conduct, specifically in the context of ethics in politics today and outlines the rise of medical narratives that provide the justifying framework for violent acts of intervention as preventive measures. I argue that such a conception not only reduces ethics to a technical practice (rendered as code and facilitated through algorithmic operations) but also puts ethics beyond contestation through its reliance on professionalism and ostensibly superior modes of technology. The result, I suggest, is an *adiaphorised* form of ethics that not only justifies but in many cases also legislates for violent interventions on the basis of a deep techno-biopolitical logic.

Considering the techno-biopolitical human subjectivities emerging from the conditioning outlined earlier, and with a view to current and future trends in military technologies, I unpick in Chapter 6 the co-constitutive relationship between biology and technology, with a focus on technological enhancement of the human and military robotics for the purpose of armed conflict. Developing the problem of the human–technology–ethics nexus, this chapter establishes how the human is rendered and understood in algorithmically analysable terms in today's practices of war and political violence. It traces human life as conceived in and as code and delineates the biopolitical ideology of the progression of the human in terms of capacity, functionality and performance. The chapter suggests that this biopolitical techno-subjectivity is shaped to fit into a biopolitical sociopolitical context of development and progress and gives rise to the functionality mandate of life. This is notable in modern forms of warfare where the human is increasingly considered to be the weakest link, framed either in terms of danger or in terms of inadequacy. I argue here that there is a notable philosophical shift in the role the human assumes in relation to technology. Finally, I argue that the resulting appearance of the human as a 'weak link' in contemporary warfare serves to further advance the legitimation and deployment of new violent technologies.

The use of armed drones for targeted killing is the most current and clear manifestation of the legitimation of these violent practices and technologies. In Chapter 7 I argue that lethal drone technology and the military practices associated with it constitute the starkest indication to date of just how deeply anchored violent technologies are within contemporary biopolitics, as well as how morally dubious it is to make claims of ethical superiority on their behalf. The chapter thus provides a detailed, real-world illustration of the influence that advanced biopolitical technologies exert on practices of war and conflict, as well as a normative critique of the biopolitically informed justifications for violence that such technologies tend to produce. I conclude the book with a call to move beyond technologically informed conceptions of ethics, and to engage more proactively with possibilities for an ethics of responsibility. In so doing I return to Arendt's notion of politics proper, arguing that its constituents – uncertainty, plurality, vulnerability – are the *sine qua non* of ethics proper, and that with a properly political form that reaches beyond biopolitical logics we might be able to restore some ethicality to contemporary ethics. This, I argue, is crucial if we wish to retain the ability and responsibility to answer the question: 'What are we doing?'

Notes

1 For consistency of terminology, however, I use the phrase 'war on terrorism' throughout the book.
2 Timothy Campbell's recent study on 'technology and biopolitics from Heidegger to Agamben' presents an insightful exception (Campbell 2011), as does Braidotti's work (2011; 2013).
3 Notably Arendt's essay 'The Conquest of Space and the Stature of Man', in *Between Past and Future* (2006b) focuses on the consequences of modern technology for the role and status of the human within a human-made world.
4 As put forward in an Oxford Union debate on the motion: 'This House believes drones are ethical and effective'. The conjoined adjectives ethical and effective are indicative of a certain conflation of the categories ethical and efficiency in the current discourses on drone warfare.
5 As Ursula Ludz highlights, Arendt was in fact very concerned with the creation of the basis for a 'new political morality' even though she never explicitly stated so out of modesty (2007: 802).

1

Biopolitics and the technological subject

> In the last resort, it is always life itself which is the supreme standard to which everything else is referred, and the interests of the individual as well as the interests of mankind are always equated with individual life or the life of the species as though it was a matter of course that life is the highest good.
>
> Hannah Arendt, *The Human Condition* (1998: 311–12)

To understand the implications of how techno-biopolitical logics shape our thinking about politics and ethics, we must first look at how this contemporary condition is produced. To do this, I look to a less obvious thinker of biopolitics, but one 'whose writings on life and politics are "biopolitical" in all but name' (Campbell and Sitze 2013b: 23). In current analyses of biopolitical structures and practices in modernity, Arendt appears again and again as a relevant theorist.[1] Like Foucault, Arendt was attuned to the modern incorporation of life and politics; unlike Foucault, her concern is not with the 'problematization of life by politics' but rather with the frustration of politics when life becomes its central theme, and scientific-technological processes its chief mechanisms (Campbell and Sitze 2013b: 25). Specifically the work of Kathrin Braun (2007), Andre Duarte (2006) and Miguel Vatter (2009), as well as Giorgio Agamben's (1998) and Roberto Esposito's (2008) analyses of biopolitics, establish some of the finer points in Arendt's exposition of the shift of life into the centre of political concerns within a wider biopolitics discourse. Although these efforts have provided an excellent point of departure to situate Arendt firmly as a thinker of biopolitics, to date, her work has remained at the margins of the biopolitical turn in scholarship, and no distinct trajectory of the elements of what constitutes biopolitical traits in her work has been established yet. This is somewhat surprising, given that Arendt was one of the early theorists who clearly understood the genesis of life as the main referent for politics in modernity and recurrently addressed the complexities involved in the modern matrix of life and politics throughout her work (Esposito 2008: 149). However, some theorists in this area see the biopolitical understanding in Arendt's work as incomplete or insufficient (Agamben 1998: 3), if not a contradiction in terms within Arendtian categories (Esposito 2008: 150). Both are fair critiques, but, rather than diminish the value of Arendt's insights into the biopolitical condition, they highlight that we must approach the topic with nuance and an open mind.

In this chapter, I unpack the biopolitics analysis in Arendt's work and show that she was both comprehensive and coherent in her understanding of the category life and its relation to politics in modernity. In doing so, I address some of the concerns raised above to establish that it is possible to place her as a relevant thinker within the wider biopolitical discourse and employ her specific perspective as a lens with which to examine the biopolitical conditioning of contemporary modernity.

Arendt herself makes no mention of the term 'biopolitics' as such, even though the expression had been used sporadically by social scientists since 1911 and within a political theory context since 1916 (Turda 2010: 112; Esposito 2008: 16–17). Nonetheless, I argue that she understood perhaps more comprehensively than other political theorists of her time, and possibly after her, the *roots* of a much broader spectrum of biopolitics than is commonly acknowledged: a form of biopolitics that is closely linked with the scientific-technological condition emerging in modernity and which has become manifest in contemporary times. In the following, I trace the historical trajectories Arendt establishes as pivotal for a modernity that is focused on life processes as a chief political aim. The chapter first discusses where in Arendt's work the shift toward biopolitics is most evident before then looking at the historical trajectories that facilitated this development and the shift in subjectivity that enables the perpetuation and growth of a modern biopolitical condition. The final part of the chapter tackles the seeming contradiction in terms implicit in bio(logical)politics, and proposes that we take into consideration a specific meaning of politics in modernity that differs from what Arendt had in mind.

Shifting life into politics

Arendt's account of the shift of biological life processes into the focus of modern politics unfolds at various points in her writings and weaves through her work as a recurring referent for her political analyses; specifically so in the texts *On Revolution*, *On Violence* and *The Origins of Totalitarianism*. However, a thorough analysis of the development and consequences of biological life as politics emerges most comprehensively in her seminal work *The Human Condition*, published in 1958. As Margaret Canovan points out, *The Human Condition* is most fruitfully read within the context of her earlier work on totalitarianism (1995: 99). And indeed, while there is no explicit exposition of life-politics in *The Origins of Totalitarianism*, her critique of the consequences of the political reduction of human life to its bare biological state is revealed at its clearest in a final chapter added to the second edition of the book, titled 'Ideology and Terror: A Novel Form of Government'.[2] Here, in her critique of Darwinian theories of teleological progressions of nature, Arendt begins to consider more carefully the implications of biological life as a political modality in modernity.

The Human Condition chronicles the different forms of activities relevant to the modern human. It was never intended to offer a systematic statement of Arendt's political philosophical thought (Canovan 1995: 99, 90). Nonetheless, it is here that she most clearly provides her readers with two crucial political analyses: a cogent account of a scientifically minded, technocratic political environment and a coherent critique of a modernity that is firmly centred on life and its processes as a political fulcrum. In this lucid 'framework for a phenomenological anthropology' (D'Entreves 2006: 56), Arendt develops a convincing genealogy of the centrality of life for politics in modernity, and the tensions this bears for the sociopolitical body in modern times. In order to better situate Arendt's account of biopolitics within her broader analysis of the conditions of modernity, it is helpful to briefly outline some key ideas and categories that serve as an organising principle for her thoughts in this text.

The Human Condition presents a complex phenomenological analysis of human activity against a tapestry of historical features, relations and stages (Canovan 1998: ix; D'Entreves 2006: 56). It is a dense work, packed with categorisations and distinctions, where several lines of thought entwine and unfold at different points in the text, offering a glimpse into Arendt's own modus of writing as a path to understanding. Arendt herself had considered it as a critical introduction to a continuing investigation into political theory, which was alas never completed and remained only a preliminary foray into the subject (Canovan 1995: 99). The compact intricacy of her framework categories (labour, work, action, earth, the world, the public, the private, the social and the political) contained in *The Human Condition* can be mitigated somewhat by untangling the work's essential organising principles. The categories labour, work and action detail human activities; the world and the earth are the loci within which these activities have relevance. The private, the public and the social constitute the realms within which the relevance of the activities is shaped. Further to this, Arendt identifies a number of basic conditions of human existence: natality (the possibility of birth and new beginnings), mortality (the futility of mortal life), plurality (the condition of multiple, inherently different humans), worldliness (the human-made surroundings upon which human existence occurs) and the earth (the physical planet) (1998: 7–9; D'Entreves 2006: 56). It is the interweaving of these categories and conditions that provides the analytical tapestry for her critique of the modern condition. In their interrelatedness, they form the basis of her concern 'with the *settings* of politics rather than politics itself' (Canovan 1995: ix, emphasis added). The complexity of this set up is indicative of the nuanced nature of her aim here: to think through the human condition anew from a different, a modern vantage point; 'to think what we are doing' (Arendt 1998: 5). Having established the core categories and conditions, she draws a historical trajectory to look how events have shaped categories and conditions, producing new, different subjectivities, which then, in turn, shape the *vita activa* in modernity.

Departing from a historical background rooted in the Greek notion of political life, Arendt understands life proper in the Aristotelian sense as comprising a narrow notion of biological life necessities (*zoe*) and a more comprehensive, public and political life (*bios*) (1998: 97). As the focus for her analysis of active life in the past and present, Arendt organises her investigations around the three strands of activities upon which life and its unfolding relies. First, 'labour' (*zoe*) is the category that is concerned solely with life processes and necessities. 'Life itself' is the human condition of labour (1998: 7). Second, 'work' (*poiesis*) is the act of fabrication, of making and creating artefacts and objects that then come to constitute a common and stable shared world for humans to exist privately and publicly. Its human condition is worldliness. Finally, 'action' (*praxis*) is the 'political activity par excellence' and requires the partaking of free and equal persons within a public sphere. The human condition of action, and thus the quintessential condition for politics, is plurality (1998: 7, 79–247). The conceptual and spatial realms for the human activities and their conditions also play a crucial role in Arendt's analysis. She draws distinctions between the private realm, the social and the public (political) sphere (1998: 22–78) to which activities and conditions have traditionally corresponded and which have provided the foundation for active human life in the past. In modernity, Arendt argues, these associations of activities, spheres and conditions no longer hold, jeopardising the possibility of an active political life in modernity. A key driver in the modern frustration of politics is for Arendt the shift of life into the centre of politics. In this delineation of activities and their respective realms we find the first key strand of Arendt's insights into this shift.

Labour, which 'assures not only individual survival, but the life of the species' (1998: 8) is, for Arendt, traditionally related exclusively to the private realm, hidden from the public at large and neither part of nor

relevant for politics proper in Arendt's understanding. Life itself, its processes, matters of birth and death are connected to this realm. Work, as
we have seen earlier, is the human activity that is required to produce a
common and shared world by creating artefacts through which people can
relate and endure. It is precisely the activity of work that can bestow 'a
measure of permanence and durability upon the futility of mortal life and
the fleeting character of human time' (1998: 8). Work connects humans to
the human-made world, to worldliness. Action, finally, is the activity associated with authentic politics, or politics proper in Arendt's account, and
can only play out in a public realm. The public realm is one of contingency
and peril, but within which humans can appear as political beings. Action
brings forth new beginnings, and thereby also comprises the Arendtian
condition of *natality* – the very condition of birth, wherein new beginnings are immanent. The political public realm is where plurality appears
an immanent condition. Its existence is, for Arendt, fundamental to the
possibility for politics proper (1998: 175).

The unravelling of tethered categories begins with a shifting of life itself
out of the obscurity of the private, non-political realm, into a public sphere.
While the distinction of *zoe* and *bios* corresponded originally to the realms
of the private family household and the political arena of the public sphere,
respectively, the shift of life into the focus of a public space has unsettled
this distinction. Moreover, this shift has given rise to the emergence of a
new, a third sphere; a hybrid sphere in which labour, work and action
eventually become undone; a realm that Arendt contentiously terms 'the
social'. Providing a sphere which subsumed concerns that had hitherto
remained within the domain of the private sphere of the household, the
modern social realm presents a quasi-public arena, situated between
private and public arenas of life. In a biopolitical modernity it is within
this realm that the individual is both depoliticised and simultaneously
becomes a subject of the functionalities of political administration as a
member of the human species (1998: 38–49). By absorbing the functions
of the family household – its concerns with birth, death, the management
of life processes for the perpetuation of the family – the social sphere
makes life processes publicly visible and politically relevant. While this
might, on the surface, seem to be an important move for the fair treatment
of humans in a contemporary liberal modernity, it presented Arendt with
a problematic obstacle to politics. The emergence of a modern society
where 'the fact of mutual dependence for the sake of life and nothing else
assumes public significance and where the activities connected with sheer
survival are permitted to appear in public' delimits the possibility for an
authentic political community enormously (1998: 46). Where life takes
centre stage in politics, Arendt saw no room for the political activities she

envisions for active, free and equal people. The prime concern for Arendt is that the rise of the social realm engenders a change in meaning of both action and politics. Action becomes behaviour and politics becomes administration. This shift is crucial for Arendt and her concerns with the settings for politics; she dedicates the entire final chapter of *The Human Condition* to this development (Canovan 1995: 248–325). For her, authentic political action is far removed from our contemporary understanding of politics as administration; this change is a lamentable but essential feature of modernity. In a biopolitical modernity, the problematic centrality of life itself takes not one but two forms. I will detail these in the remainder of the chapter. Arendt's conception of 'the social' as distinct from 'the political' is a controversial issue, and a number of scholars have noted that it remains ill-defined as a sphere (Fenichel-Pitkin 1998; Owens 2012). However, it is in understanding the social in a biopolitical context that some of Arendt's contentious assertions regarding this quasi-public realm might become a little less jarring.[3]

In *The Human Condition*, Arendt traces a historical trajectory of the central factors that gave rise to a modern life-politics perspective back to a number of key developments in history, beginning with the late sixteenth century. These include the Reformation and the resulting expropriations, the 'discovery of the Archimedean point', the advent and predominance of science and technology, and the general 'shrinking' of the earth as a shared space. In intricately related ways, each one of these historical turns has contributed to generating a modernity in which important boundaries of categories and distinctions have all but disintegrated and have given way instead to one dominant interest of society: that of its own perpetual survival, reproduction and progress, facilitated by ever more autonomous technological machines that may well subsume both life and political perspectives more fully in the not-too-distant future (Arendt 1998: 2). If we want to bring in Arendt further into the fold of biopolitical analyses, it is useful to take her tethered categories and their apparent disintegration seriously. To more fully understand her strong grasp of biopolitics, a broader perspective is required to discern the nuances with which she looks at the life–politics relation.

Throughout her writing, Arendt offers a critical perspective that considers the prevalence of life at the centre of politics in not one but two significant aspects: life as the very target of political practices and their thanato-political consequences on one hand, and the theory of organic life processes and biology as serving as the very basis for political practices in modernity on the other. As such, her account considers the *making-political* of biological life processes (*zoe*), emphasising the organic dimensions of the human and the political body as a living being in pursuit of one supreme goal: the perpetuation of the species. Yet Arendt reaches further in her analysis of life-politics and ties in an ancillary aspect, in which she problematises the scientific process of truth-finding in a society in which *zoe* dominates the social realm and political concerns, and which has a different biopolitical manifestation: the possibility for a naturalistic conception of political ends. The former approach has dominated scholarship on biopolitics in political theory in the past decade, and broadly reflects Michel Foucault's expositions of power relationships and biopolitical *dispositifs* on the subject. In contrast, the latter – theorising the allegorical underpinnings of organic and biological processes for the functions of the (nation) state – is often overlooked as a relevant facet of contemporary biopolitics. This is, however, a useful aspect, I argue, because it ties in more clearly the prominent role of technology, biology and politics in contemporary modernity and is instrumental in understanding the turn towards violent solutions for political problems. I discuss this more thoroughly in Chapter 4. For the analyses in this book, I look at both interlinked aspects of modern biopolitics: the politicisation of *zoe*, where biological life becomes a political target, and the *zoe*fication of the political, which comprises the conception of the state and its people as a living organism that adheres to natural, biological laws and is imbued with physical qualities that can be scientifically ascertained, and technologically adjusted. Both have an impact on the production of contemporary (techno-) subjectivities and both determine political practices accordingly. Considering both of these aspects of biopolitics therefore provides us with a more comprehensive lens with which to examine political practices, mechanisms and activities and their relation to and ethical reasoning for technologically mediated political violence in modernity.

The next section establishes the dual aspects of biopolitics in Arendt's account, and engages with the genesis of these two strands of life politics to show how they form the backbone of a biopolitical analysis in Arendt's work. I begin by unpacking the implications of a key event that provided the material foundation for the manifestation of biopolitics for Arendt: expropriation and consequently world alienation.

Politicisation of *zoe*, *zoe*fication of politics

The concept of world alienation captures for Arendt the historical emergence of a modern economy that prioritised wealth creation over property ownership, a process that was set off by a wave of expropriations during the Reformation (1998: 254–5; Canovan 1995: 150). World alienation, for Arendt, refers to the ordained unsettling of individuals and families from their property, and with it from their conditioning link to the world as a place of permanence and stability – an important source for ontological security prior to the Reformation. Recall that, for Arendt, the existence of artefacts provides a crucial link to the condition of worldliness. The aliena- tion engendered by expropriations is thus not merely of material nature, but also produces a changed existential lens. I will come back to this point in more detail further on. The expropriation of certain classes (initially the poor, but this was later extended to other classes as well), and the severance of their ties to what hitherto had been understood to be a common and shared world through property as a specific place for each family and its household, generated a growing population whose property was turned into capital, and whose bodies were turned into labour (Canovan 1995: 150). The wave of expropriation was pivotal in initiating the liberation of the labour process from the previously secluded private sphere of the household and moved it into the public sphere of a society. It is on the basis of this development that Arendt claims a hitherto non- existent, different class of labourers had emerged who 'stood not only directly under the compelling urgency of life's necessity but was at the same time alienated from all cares and worries which did not immediately follow from the life process itself' (Arendt 1998: 255). Stripped of property and thereby alienated from the hitherto shared world, the poor and the peasantry, and with advanced modern societies ever greater segments of the population, transformed into a mere manifestation of biological life processes, which could then be managed as assets. Margaret Canovan elaborates this reshaping process:

> When peasants were deprived of the property that had given them their place in the common world, and were reduced to day labourers, they were, according to Arendt, transformed into embodiments of mere biological processes. (1995: 83)

In this process, the labourer emerged from the 'hidden' private realm of the household, now appearing within a public social sphere, and at once not only gained significance as a resource for the cyclical life process of production and consumption, but was made into an element of the collec- tive species 'human', whose survival is now publicly organised within and by the functions of the social. The newly revealed prevalence of labour as an activity in the public domain then required the public management of labour's condition: biological life itself. Now that labour was shifted into the publish sphere, as an integral element for the modern political economy, life itself, its concerns with birth, death and survival were made into a collective concern. It is within this collective public sphere that the social realm has now taken on the mandate to ensure the survival of the species and the very nature of the labouring process. Here, the taking care of life's necessities and the meeting of biological demands gains prime economic and political significance. It is in the liberation of labour power and the release of biological processes into the public sphere that we see two crucial biopolitical developments in modernity: the production of the public human first and foremost as a member of the species 'human', and the formation of a political economy in which biological life processes form the model for the modern production of wealth creation and capital accumulation. What emerges here is a model of wealth accumulation which is 'stimulated by the life process and in turn stimulat[es] human life', where labour power fuels the production of surplus from the cycles of production and consumption (1998: 256). Not only is life as such politicised by being fed into the public production of wealth for society, but life as such and the logic of its processes serve as the model for the modern political economy for the 'life process of society' (1998: 255). Instrumental to this is the growing dominance of science and the discovery of scientific processes and process-thinking. As Arendt observes,

> [t]he coincidence of Marx's labour philosophy with the evolution and development theories of the nineteenth century – the natural evolution of a single life process from the lowest forms of organic life to the emergence of the human animal and the historical development of a life process of mankind as a whole is striking. (1998: 116)

What was common to the development theories in the early sciences is the idea of both process *and* progress. Both the natural and the social sciences pursued generalised theories of processes for the progressive evolution, or development, of events. Once these processes had been established, they could be technologically reproduced or intervened in to safeguard progress. It is for Arendt not surprising then that the identification of the biological process should consequently serve as the logical map for the life processes of society through labour power.

World alienation, and the consequences associated therewith, thus reflect the severance of individuals' relationship with a shared and common world, which ensued the moment when persons were deprived of the property that both sheltered them from and simultaneously connected them to a common world, and when those functions belonging to the private realm – matters of *zoe* – collapsed into the public realm (1998: 254–5). The function and role of a common and shared world that lends permanence to human life is of crucial significance to Arendt. As we have seen, the Arendtian category 'the world' is a product of the human aspect of 'work', it is created not solely for immediacy or consumption but for providing a lasting temporal structure that exceeds the biological life cycle. It is this human-made world that gives the human a context upon which to act and upon which meaning is bestowed. The world, as a tangible stage where human histories and history unfolds, connects the plurality of humans, while simultaneously holding space to ensure the freedom to engage in political action. For Arendt, 'without the world into which men are born and from which they die, there would be nothing but changeless eternal recurrences, the deathless everlastingness of the human as of all other animal species' (1998: 97). Only the existence of a shared world ensures transcendence; only by bestowing a narrative structure on to life can existences achieve meaning, such that events can come to pass, and be passed on, as histories in and for a shared, lasting world. A shared world serves as a fundament for political acts. It is through the alienation from this shared and common world that the reduction of the human to his biological functions can ensue, in a sacrificial turn. As Arendt notes, 'the process of wealth accumulation, as we know it, ... is possible only if the world and the very worldliness of man are sacrificed' (1998: 256).

In this alienating shift, the creation of wealth for the ongoing construction of, and contribution towards, a shared and permanent world gave way to a *zoe*fication process in which the very basis for the broad category of life, one that allows for a condition of plurality among humans, became drastically reduced to the necessity of life functions. Instead, an economy emerged that *mirrors* the life processes, channelling the 'fertility of natural processes' in the creation of a surplus that feeds back into a perpetual cycles of creation and destruction (1998: 255). As Arendt argues, 'under modern conditions, not destruction but conservation spells ruin because the very durability of conserved objects is the greatest impediment to the turnover process, whose constant gain in speed is the only constancy left where it has taken hold' (1998: 253). In other words, in a *zoe*fied public sphere, permanence is a liability, akin to what Zygmunt Bauman identifies as *Liquid Life*, the 'consuming life' that 'casts the world and all its animate and inanimate fragments as objects of consumption' (Bauman 2007: 9). These objects of consumption have a limited shelf life and an inherent mandate for constant renewal. In such a form of society, the work aspect of the human condition, so crucial for the creation of a shared world fundament, is rendered solely instrumental, whereby the products of work are no longer the building blocks for a common lasting world in which humans can present their unique selves; work is subsumed by labour. In other words, the essence of *homo faber* – the maker of artefacts for a shared world stage – becomes frustrated. Instead, Arendt argues, the instrumentalisation of work as process renders *homo faber* essentially a labourer (1998: 305–8). Biological life processes of production and consumption have then not only become the prime preoccupation of human productivity, they have indeed become the very underpinnings of the understanding of humans within their environment. Arendt recalls a Marxian expression for this transformation, the '"life process of society" ... whose wealth-producing capacity can be compared only with the fertility of natural processes where the creation of one man and one woman would suffice to produce by multiplication any given number of human beings' (1998: 255). In such a society, humans as such fulfil primarily a functional mandate, and, eventually, come to see themselves as such, within the wider body of a biopolitical community.

Where functionality becomes a *raison d'être* for human existence, plurality becomes an obstacle. This engenders the blurring of boundaries between the private and the public, replacing, through the emergence of the social, the family as the subject of life processes. In this absorption of family within society, '"blood and soil" is supposed to rule the relationship between its members, [and] homogeneity of population and its rootedness in the soil of a given territory become the requisites for the nation-state everywhere' (1998: 253). Where homogeneity is required, plurality is diminished and the *conditio per quam* of political life is imperilled (1998: 7). The social then subsumes the political, whereby political action, in its delimited form, has given 'place here to pure administration' of life necessities (1998: 45). But how did this overarching biological consciousness emerge? And why did this urge of life necessities enter the public realm?

The development of a modernity that is focused on life functions as a political concern is brought to a conclusion in *The Human Condition* with what Arendt calls *'the victory of the animal laborans'* (1998: 320). Here, she delineates the emergence of a self-understanding of man as a scientific, biological being, not just in the private domain of the household but rather on the public stage. I will take up the relevance of this scientific turn in the next section. It is in this final section of *The Human Condition* that she emphasises the transformation of subjectivity in a biopolitical modern society most clearly (1998: 321, 322). Where, by way of world alienation, the modern process of secularisation, and the loss of 'certainty for a world to come ... the only contents left were appetites and desires, the senseless urges of the body' (1998: 320–1), death and mortality became an existentialist quandary, and the only conceivable possibility for immortality was presented by the sheer cyclicality of the life processes. Hitherto, permanence and transcendence were bestowed by a shared world as a political stage, and the spiritual idea of an afterlife. Now the fluidity of the modern world gave the condition of mortality an uncomfortable existential finality and a peculiar dominance to life itself as a condition of labour. In Arendt's words:

> The only thing that could now potentially be immortal, as immortal as the body politic in antiquity and as individual life during the Middle Ages, was life itself, that is, the possibly everlasting life process of the species mankind. (1998: 320–1)

In the modern sociopolitical body, where a society has but one chief and common interest, that of the survival and progress of the species, individual life is placed under the mandate of this goal. Instilled, however, in this form of society is the acceptance of the priority of the biological, made manifest as a demand within a public realm, whereby this type of society of labourers 'demands of its members a sheer automatic functioning, as though individual life had actually been submerged in the over-all life process of the species' (1998: 322). Again, the emergence of the social realm is crucial here.

This formation at the heart of the *zoe*fication of society has broad political consequences. The shift of life necessities and biological concerns from the private realm of the household and family into the social sphere and further, in the modern age, into the opaque nexus of the social and the political reflects a growing predominance of biological concerns as a more widely accepted public matter. As Arendt observes, '[t]hrough society, it is the life process itself which in one form or another has been channelled into the public realm' (1998: 45). Where previously the survival not only of the individual but also of the species was secured and guaranteed by the family, the mandate of this function has, according to Arendt, become a public matter in the social sphere. Hanna Fenichel-Pitkin gets to grips with this vague domain in her work on Arendt's realm of the social, and the coupling of life processes with this sphere is not lost on her. She notes: 'the social connects with labour, since both the productive labour that supplies food and shelter and the reproductive labour by which women give birth are "subject to the same urgency of life" that rules over "the realm of the social"' (Fenichel-Pitkin 1998: 187). Arendt terms this the 'unnatural growth of the natural' in society, a development she is strongly critical of (Arendt 1998: 46).

This social realm thus is 'the *form* in which the fact of mutual dependence for the sake of life and *nothing else* assumes public significance and where the activities connected with sheer survival are permitted to appear in public' as political processes (1998: 46, emphasis added). It is the life of the 'monolithic' collective, not of the individual, that is at stake here in this biopolitical process. The logic of life processes as the prime organising principle for a sociopolitical community, however, means that membership in the community can, but need not necessarily, be delimited by 'blood and soil'. Recognising the potential for a wider expansion of this bio-logic, Arendt foreshadows the replacement of a territorially bound society with a notion of a comprehensive humankind, and that of state property with the entire earth, just as 'family' had previously been replaced by 'nation' in the pursuit of perpetuating and progressing life at large (1998: 257). And it is in this turn that Arendt sees the dangers of a 'worldless mentality of ideological mass movements', producing demarcation lines of inclusion and exclusion, based on fitness for the life processes of humankind (1998: 257).

Prioritising life processes

What does this development mean for the relationship between life and politics, specifically? And how does this relate to what Arendt calls the social? What is essentially at stake here in this twofold biopolitical dimension is, on one hand, the transference of the care and concern over life processes into the social realm, where societal homogenisation and political administration are prevailing factors. And, on the other, the rendering 'biological' of political processes, in the continual creation of surplus and value through the reproduction of life processes, for the survival and progression of the species, perpetuated in an endless cycle. In this condition, life as *zoe* is both target and foundation for a political perspective upon which administrative measures are built in the modern age. The other facet of life, *bios*, and with it the capacity for authentic political action in the Arendtian sense is rendered virtually incapacitated in the social realm. Fenichel-Pitkin recognises the biological relevance of this constellation clearly, even though she makes no reference to the biopolitical dimension here. For Fenichel-Pitkin, it is evident that the growing predominance of the 'natural' is problematised when 'politics ... would supply the collective body with a head to reassert human direction of its biological processes, of the socioeconomic forces generated by large numbers of interdependent people making their living' (1998: 11). What she perhaps inadvertently identifies here is that the allusion to the political collective as a biological body serves not merely as a metaphor but as a performative concept, or plan, for a new understanding of politics.

For Arendt, the troublesome tension in the public sphere, which comprises both the social *and* the political equally, arises precisely if either the capacity for action is compromised or the condition of plurality is jeopardised. She locates the manifestation of the blending of the social and the political in a specific historical event: the French Revolution. In *On Revolution*, Arendt illustrates the fateful collapse of the private and public sphere with the storming of the *Sans Culottes* on the political scene, demanding for their needs to be met, thereby giving rise to the political centrality of life struggles, or poverty (Honig 1995). Arendt succinctly, and critically, observes that when the poor 'appeared on the scene of politics, necessity appeared with them' (2006a: 50). And with a sudden prominence and pre-eminence of life processes in the public realm, matters concerning *zoe* subsumed a space held for the creation of freedom for political action. Furthermore, right was constituted primarily as a right to food, drink and the right to reproduction – in short, the right to perpetuate not just individual life but rather that of the species, that of humanity. Not only had matters of life's necessity for survival taken political centre stage, but concomitant with this was the emergence of a spectral political collective body that voiced the joint and sole political demand for the fair administration of the materials and conditions for life's necessities. The conflict this creates is obvious when we remember Arendt's claim: 'in politics not life but the world is at stake' (2006b: 155). And it is precisely the jeopardy this world, the public space of plurality, is in, when life, in its biological, cyclical sameness of life processes, becomes the core of politics, subsuming a heterogeneous plurality (of humans who each carry unique stories, attributes and qualities) into one singular concern of 'a mass that moves as one body and acts as though possessed by one will' (Arendt 2006a: 84). In other words, the political basis is lost when political matters acquired a

> biological imagery which underlies and pervades the organic and social theories of history, which all have in common that they see a multitude – the factual plurality of a nation or a people or society – in the image of one supernatural body driven by one superhuman, irresistible 'general will'. (2006a: 50)

This idea of a monolithic public space that is occupied by one dominant common interest, that of the survival and progression of the species, enforced by 'sheer number' (Arendt 1998: 40), does not sit comfortably with Arendt. Here she becomes provocatively strict in her distinctions as she insists that what develops among humans in the social realm represents the antithesis of what she values in the human condition for political action. The full development of the social and the absorption of all life functions into the heart of its remit means an almost inevitable exclusion of the possibility of action and therefore of politics proper all together (1998: 40, 45, 321). Just as action is excluded from the original realm of the private household, as conditions of freedom and plurality cannot be sufficiently met, the social, as a kind of expansion of the household and its functions, offers no space, or even the need for action, in order to fulfil its mandate. This delimits the possibility for politics in the Arendtian sense significantly. In Arendt's words: 'it's decisive that society, on all its levels, excludes the possibility of action' (1998: 40). What replaces action is 'a certain kind of behaviour, imposing innumerable and various rules, all of which tend to "normalize" its members' (1998: 40). Recall that action is for Arendt the category where political acts and political agency reside. The frustration of this category puts the possibility for politics in peril.

The urgent priority the social places on biological necessity jars strongly with the ideal Arendt has developed of a public space in which it is the plurality of humans of difference that allows for the reality of action. Laura Bazzicalupo summarises Arendt's objections succinctly: 'When the cyclical metabolism of life occupies the entire public space in which the circle of production and consumption cannot be broken, we are ruled by necessity and survival, which is neither human nor politically free' (Bazzicalupo 2006: 112). In other words, when the life and survival of the species is at stake, no form of politics (proper) is given. For government, this means an exclusive twofold mandate: the administration of the means to perpetuate the life processes of society, and providing security for the survival of the population.

For Arendt, the reduction of the possibility of action also represents a stark infringement of the possibility of freedom. She explains the detriment this has for freedom in modernity in her essay 'What Is Freedom'. Again, here, the scientific turn plays a key role:

> The rise of the political and social sciences in the nineteenth and twentieth centuries has even widened the breach between freedom and politics; for government, which since the beginning of the modern age had been identified with the total domain of the political, was now considered to be the appointed protector not so much of freedom as of the life process, the interests of society and its individuals. (2003b: 443)

For Arendt, the thought that the securing of life processes could result in greater levels of freedom is a distortion of the very categories 'life' and 'freedom'. In Arendt's analysis, the life process 'is not bound up with freedom, but follows its own inherent necessity' (2003b: 443). In this view, freedom is entirely unrelated to the very character of life processes and their immanent urgencies. Consequently, the political securitisation of the life processes for the creation of freedom is a paradox in so far as it applies only to the very margins in which life itself is at stake.

We have seen so far that the social sphere is the space in which humans in modernity not only lose their capacity to act but also lose the material and physical space associated with it. The reduction of the human to a member of a species always represents for Arendt a diminution of their full human capacities. In the social sphere, which finds its *raison d'être* in the maintenance of the life processes of the species, these limitations become manifest. In this realm, the common denominator of togetherness is not realised by the ability to act on difference, or within plurality, but rather based on the biological functions of the species, which requires the homogenisation of the group for its benefit. This, in turn, engenders conformism, thus paving the way for the replacement of contingent action with predictable behaviour in a normalising turn. Less focused on the specific mechanisms that facilitate the normalisation as such, Arendt here is concerned with the conditions this provides for the modern human.

It is in this aspect that Seyla Benhabib levies her strongest criticism against Arendt's analysis. She finds this element of the social in Arendt lacking in insight, and critiques that Arendt not only avoids a discussion of the logic of rationality that renders behaviour predictable and homogenous, as Max Weber offers in his writings on *Sozialpolitik*, for example, but also fails to engage in a deeper investigation of the institutions that facilitate the normalisation process. As such, Benhabib finds Arendt's exposition of the social as a contribution to social theory 'thin and sometimes reductionist' (Benhabib 1996: 26) and concludes that this element of Arendt's work must be read against the background of, and with, other twentieth-century writers to make her work less implausible. Recall, however, that Arendt's main concern is an analysis of the settings for the human condition. This is why she reaches far back in her analysis and perhaps does not give the modern institutions the attention other scholars have given them in analysing the distinctly modern sphere of the social. I would, however, argue that Arendt has a very thorough and clear understanding of the specific rational subjectivities that were produced in biopolitical modernity. Moreover, in her analysis, Benhabib avoids a discussion of the life processes that so completely dominate the social sphere and she takes for granted the merger of an economisation of life with the rise of life necessities in the public sphere, without giving any further attention to the stark shift of focus on biological life in the social. It may be this inadvertent 'blind-spot' with regard to the biopolitical dimensions of the social realm that gives rise to Benhabib's claim that Arendt's analysis of mass society is left wanting in terms of what it reveals about mechanisms and normalisation practices. When read against a biopolitical background, it becomes evident that Arendt's work offers a different but compatible perspective of how mass behaviour in a biopolitical society is normalised. Again, her analysis is less concerned with institutional mechanisms of the social, but rather focuses on the historical developments that have had a lasting influence on altering political perspectives and subjectivities, and which provide the cartography for a set of mechanisms for normalisation in the modern human condition. As such, the developments that have enabled the shift of life into the centre of politics all contribute to the emergence of normalising mechanisms, whereby the rise of biologically infused political processes serves as the master mechanism for political administration. Rather than offering an analysis of the mechanisms and institutions, she offers an analysis of the genesis of the mechanisms and institutions in biopolitical modernity and, importantly, an astute insight into the advent of the specifically modern condition that facilitates and sustains a biopolitical reality: the scientific-technological mindset.

The birth of the technological human

In the final chapter of the *The Human Condition*, 'Vita Activa and the Modern Age', Arendt charts the implications of the historical events leading to the modern human condition. The biopolitical consequences of world alienation, discussed above, present one strand of this trajectory; another important aspect is that of a shift of perspective from which modern humans were able to interpret not only their broader environment, but also themselves within it. It is this strand that represents the very underpinnings of how the institutionalisation of biopolitical life became a reality. Tracing the modern prevalence of science and technology in a political context all the way back to Galileo, Arendt paints a gloomy picture of how the shift from humans as earth-bound beings to humans as universal beings has come to facilitate and perpetuate conformity over difference, behaviour over action, codification over speech, and has facilitated a perspective that posits the human as maker of all things, including life, in modernity. The chapter abounds with a palpable disquietude over what attributes we might have lost in our capacity as humans in the modern age.

At the very heart of the advent of scientific-technological subjectivities, Arendt argues, stands the birth of algebra and a shift from the spatially bound geometrics of ancient times to the ability to 'grasp in symbols those dimensions and concepts which at most had been thought of as negations, and hence limitations, of the mind, because their immensity seemed to transcend the mind of mere mortals' (1998: 265). With this revelation came the notion that the human senses may well be too limited to reveal reality, as the 'modern *reductio scientiae ad mathematicam* has overruled the testimony of nature as witnessed at close range by human senses' (1998: 256). What ensued was a lasting reliance on instruments and implements of measurement and verification – technologies which exceed our own sense perceptions but continue to remain a human-made invention and thus originate always within humans themselves. This newfound technological dependency in turn led to a crucial change in our relationship with nature, as well as our own selves, as Arendt explains:

> even more significant was the fact that the new mental instrument ... opened the way for an altogether novel mode of meeting and approaching nature in the experiment. In the experiment, man realised his newly won freedom from the shackles of earth-bound experiences; instead of observing natural phenomena as they were given to him, he placed nature under the conditions of his own mind. (1998: 265)

With the application of abstract mathematical principles and the rising confidence in the outcome of scientific experiments, life itself, its biological mechanisms were made translatable into patterns and processes. Consequently, the mandate of the mathematisation of life, the duress of rendering life and its attributes analysable, organised and expressed in mathematical terms, and the ensuing aim to gain control over the unfolding of life processes came to shape our mortal place in the world, as part of a scientific and increasingly technologised environment, whereby 'the reading of an instrument seems to have won a victory over both the mind and the senses' (1998: 275). As a consequence, modern humans understand themselves as part of the very pattern they project themselves to be part of through their self-conceived experiments. Implied in this is a certain solipsistic logic: if all that is true is authenticated through experiments, and all experiments are conceived by the human mind, in all its limits, the modern human has little choice but to become a self-referential result of what science can demonstrate as possible. In this solipsism, we become analysable beings in so far as our biological make-up and capacities follow certain patterns that are valid within specified parameters of what constitutes humans and the potentialities they hold. In a modern society in which a multitude is present as 'sheer number' only and where the 'communistic fiction' of one shared and common goal of society, namely that of the survival and progress of the species, is prevalent, the potential for instituted normalisation processes are evident (1998: 43, 256).

Where sense perceptions no longer provide the basis for our place in the world, but must rely on the substitutive authority of technological means, the certainty of one's actions, in the private as well as the public realm, is called into question. Arendt rightly detects a reductionist aspect here:

> With the rise of modernity, mathematics does not simply enlarge its contents or reach out into the infinite to become applicable to the immensity of an infinite and infinitely growing, expanding universe, but ceases to be concerned with appearances at all. It is no longer the beginning of philosophy, of the 'science' of Being in its true appearance, but becomes instead the science of the structure of the mind. (1998: 266)

And, importantly, by referring to Marx, she notes: 'If Being and Appearance part company forever, and this – as Marx once remarked – is indeed the basic assumption of all modern science, then there is nothing left to be taken upon faith' (1998: 275). Arendt presciently anticipated the technological pervasion that has come to determine our contemporary context, and that becomes increasingly relevant in contemporary discussions on post-humanism, when she notes that 'we are surrounded by machines whose doings we cannot comprehend although we have devised and constructed them' (Arendt 2006b: 264). When sense perceptions are insufficient in grasping the world, everything must be doubted, even to the extent of one's own authority over biological functions and one's political role. With the science revolution, truth comes to be tested and verified through hypotheses. And through these essentially human-made processes of verification, we have acquired the ability to analyse, systematise and eventually control nature. The paramount points of authority in Arendt's assessment of this *reductio scientiae ad mathematicam* are the schemata of the human mind as such, providing a reference point for what is real and certain as 'a framework of mathematical formulas *which are its own product*' (Arendt 1998: 284, emphasis added). Human-made verifications of mathematical logic then replace the previously established validity of and reliance on sense perceptions. In this, however, lies again a solipsistic turn that Arendt clearly identifies:

> While technology demonstrates the 'truth' of modern science's most abstract concepts, it demonstrates no more than that man can always apply the results of his mind, that no matter which system he uses for the explanation of natural phenomena, he will always be able to adopt it as a guiding principle for making and acting. (1998: 287)

This constellation, the scientific rule of a mathematical logos and the operations required by scientific inquiry provided the very basis for the primacy of labouring as a modern model for living. With the historical coincidence of the realisation of scientific processes as a means for knowledge-finding and the shift toward introspection in modern philosophy 'it is only natural that the biological process within ourselves should eventually become the very model of our new concept' (1998: 116). This turn is crucial.

Based on such anthropocentric patterns of investigation, the human mind and body become both subject and object of science and technology. It is against the application of this development in modernity and with regard to the *zoe*fication of economic and political processes that Arendt levels a grave critique at the rise of society and its preoccupation with biological life processes:

> it may be well to recall that its initial science of economics, which substitutes patterns of behaviour only in this rather limited field of human activity, was finally followed by the all-comprehensive pretension of the social sciences which, as 'behavioral sciences' aim to reduce man as a whole, in all his activities, to the level of a conditioned and behaving animal. (1998: 45)

For Arendt, the emergence of the science of 'man' as such determines the mandate of conformity in society, from which the possibility of a science of economics as ethically and politically relevant could then emerge. Again, Arendt's words are instructive:

> Economics – until the modern age a not too important part of ethics and politics and based on the assumption that men act with respect to their economic activities as they act in every other respect – could achieve a scientific character only when men had become social beings and unanimously follow certain patterns of behavior, so that those who did not keep the rules could be considered to be asocial or abnormal. (1998: 42)

In this, Arendt detects a perilous development: where biological life processes rule supreme, where scientific inquiry into the human being reduces humans to their measurable, calculable, experimentally verifiable and, not least, predictable functions, and where technology assists in further marginalising the realm of human experience, life as a broad category inescapably becomes narrowly defined as functional, biological life. In turn, the human is publicly relevant primarily as members of the animal species 'human', whose biology and behaviour can be determined, predicted and eventually controlled scientifically and technologically. In the modern context, Arendt critiques, this preoccupation with life processes – not least as the technological possibility for replication and intervention into *zoe* – produces a perspective by which life becomes a matter of production and intervention. This risks the loss of a heterogeneous plurality in a mass of sameness and cyclicality in which increased emphasis is placed on scientific-technologically ascertainable qualities. This, in turn, further delimits the possibility for action. Etienne Balibar lucidly elaborates this turn in Arendt's work and agrees that an increasingly technicised reproduction of life provides a progressively encroaching substitution of 'the good life' with alienated artificiality. The contradictory outcome of this is 'that the political becomes reduced to its natural basis in the inverted form of an absolute artificiality' (Balibar 2007: 729). Here again we see not only the realm of freedom for politics proper severely reduced, but also the possibility for action, as the basis for an Arendtian understanding of politics, considerably limited. The normalising processes that are engendered through the universality of what Arendt terms *animal laborans* – the labouring human – culminate eventually in a fully developed labouring society. Her gloomy assessment of the late stages of this process is telling. In the final phases of producing a society of labourers and jobholders, 'the only active decision still required of the individual were to let go, so to speak, to abandon his individuality, the still individually sensed pain and trouble of living, and acquiesce in a dazed "tranquilized," functional type of behaviour' (Arendt 1998: 322).

Arendt recognised the political dangers inherent in the mathematisation of life early. Foucault, only a decade later, describes this very development in terms of biopower, the power that renders the 'seething mass, which is sometimes murky and sometimes bloody' organised and manageable through the systematic collection of knowledge and hard facts (Foucault 2004: 54). Here, in Arendt's analysis of the scientific revolution and its problematic political consequences, she offers a foundational perspective for Foucault's theory of biopolitical institutions. The shift in governmental perspective from rulership to administration and the management of life functions for the sake of the species, and the demand for behaviour so as to make life statistically determinable and yield 'correct predictions' forms the very basis of the mechanisms available to establish a politicisation of *zoe* and perpetuate a biopolitical modernity (Arendt 1998: 43). This condition eventually provided an additional, pivotal sociopolitical opportunity: that for 'the society of scientists' to organise themselves in groups and bring about 'the radical change of mind of all modern men which became a politically demonstrable reality' in the 20th century (Arendt 1998: 271). Only through the behavioural integration of humans as predictable and controllable entities, as a 'mere part of a society that conditioned and determined the individuals as the whole determines its parts' (Arendt 1953: 304), and as considering themselves as self-constituting, did the possibility for totalitarian ideologies and politics become a realistic possibility. Arendt traces this most clearly in the final chapter of *The Origins of Totalitarianism*.

This crucial combination of the human as an analysable and mathematically ascertainable being, a delimited realm for political action and the concept of a 'zoefied', natural, basis for political action all have a contributory role to play in the possibility of totalitarian reign. It is where the politicisation of *zoe* and the *zoefication* of politics come together most fatefully in a biopolitical reality. It is particularly the latter concept – the naturalisation of the political – that has been identified by André Duarte to hold the strongest potential for biopolitical violence, although he does not explicitly elaborate the *zoe*fication of political acts (Duarte 2006: 418). Here, Duarte rightly stresses the problematic impact a naturalised concept of politics has on the potential for discriminatory and racialised violence. He notes: 'when power and violence are interpreted in terms of biological metaphors this can only produce more violence, especially when race is involved' (2006: 418). I discuss this relationship in greater detail in Chapters 3 and 4.

Arendt's theories of a modernity that has placed the glorification of biological processes at its political centre comprises a simultaneous critique of both Marx and Darwin. In her controversial criticism of Marx's rendering of history as a scientific process and, likewise, Darwin's theories of biology as a process, she explains how the understanding of nature or of history as a progressive and teleological development was so dangerous as a basis for totalitarian ideology. In her essay 'Ideology and Terror', Arendt critically exposes the modern change in perspective, which she ascribes to Marx and Darwin. With this scientific shift, both history and nature were no longer understood as the 'stabilizing source of authority for the actions of mortal men' but rather are defined in terms of progressive processes, in other words, in terms of directional movement (Arendt 1953: 309). Arendt understands the Darwinian concept of the development of the species as subsuming the cyclicality of biological life processes within the logic of a progressive, linear history, thereby thrusting the circularity of life processes into a development-driven linearity of life-*progress* with, ideally, predictable outcomes. She critiques Darwin strongly for this turn:

> Darwin's introduction of the concept of development into nature, his insistence that, at least in the field of biology, natural movement is not circular but unilinear, moving in an infinitely progressing direction, means in fact that nature is, as it were, being swept into history, that natural life is considered to be historical. (1953: 309)

In other words, Darwin's theories of the survival of the fittest have enabled an ideology of a natural law of movement, which appears to determine that there is a teleological direction in the laws of nature by which it is destined to fulfil itself (Villa 1999: 18). Against such a deterministic biological backdrop, what is at stake in Nazi totalitarianism as an ideology then is the fulfilment of humankind – a process that ends only when all that is unfit to live, all that disrupts the progressive process of the movement of natural laws, is eliminated fully, whereby '[c]ontingency would be eliminated, replaced by an inescapable historical or natural necessity' (1999: 20–1). In this particular political ideology, humankind is thus understood as a multitude – a multitude not in terms of difference but in its homogeneity, whereby it would become entirely feasible to identify elements that are detrimental to the species overall, and which must be intervened in or eliminated. In Arendt's analysis, in Nazi totalitarianism, this logic embraces, in principle and ambition, the whole population, the members of which, in their mandate for inclusion, are pressed against each other to become one entity. The individual quite simply becomes a member of the species. Nazi biopolitics then receives its murderous legitimacy not by the exception, as Agamben might suggest, but by referring to something with greater and more permanent legitimacy – not the laws of nature but the laws of the *development* of nature. To 'realise' humankind's potential, that which is unfit to live must quite naturally be eliminated. And this process, in theory, ends only when humankind is fully realised. It is evident for Arendt, that this development can be achieved only when the human perceives herself as part of the wider humankind and not as a unique individual, bestowed with the freedom to act. In short, this can be achieved only when the radical uncertainties that thinking and acting can engender are problematised, reduced, or eliminated. The mechanisms employed in Nazi politics to ensure that nature can 'race freely through mankind, unhindered by any spontaneous human action' are realised, Arendt suggests, through terror (Arendt 1953: 310).

> Terror as the execution of a law of movement whose ultimate goal is not the welfare of men or the interest of one man but the fabrication of mankind, eliminates individuals for the sake of the species, sacrifices the 'parts' for the sake of the 'whole'. (1953: 311)

Indeed it is through this substitution in totalitarianism that the spaces between humans, which provide the grounds not only for a condition of plurality among humans, but also for action and contingency, are eliminated, and humans in their individual uniqueness 'disappear into One Man of gigantic dimensions' (1953: 312). The 'iron band' of terror demands that the human becomes merely a part of this 'supra-human one' and is led to act accordingly. Arendt states:

> In the iron band of terror, which destroys the plurality of men and makes out of the many the One who unfailingly will act as though he himself were part of the course of History or Nature, a device has been found not only to liberate the historical and natural forces but to accelerate them to a speed they never would reach if left to themselves. (1953: 313)

In order to integrate the full plurality of multiple humans into a single entity that is concerned primarily with the development of nature, action is a hindrance and must be substituted for behaviour. While Arendt does not elaborate in greater detail the very technologies that turn action into behaviour in a society that is conceived as 'one' in *Origins of Totalitarianism,* reading this earlier text against her expositions in *The Human Condition* makes for a more complete picture. This, then, allows us to more fully comprehend how thoroughly she understood the modern preoccupation with life processes at the heart of political concerns and its grave biopolitical consequences in modernity.

This chapter has situated Arendt as a biopolitical thinker of considerable nuance and highlights the prescient analysis she presents of how life could become the central aim of politics. In the following chapters, I unpack her contribution and the implications of such a biopolitically conditioned modernity for the possibility for politics and violence in greater detail. But before I can do this, I must address what some might consider a possible problem in seeing Arendt as a biopolitical thinker. The antithetical condition of biological factors for politics proper.

The elephant in the room: Is an Arendtian biopolitics possible?

Both Roberto Esposito's and Giorgio Agamben's expositions on the subject of biopolitics are to some extent directly reliant on Arendt's work in *The Human Condition* and *The Origins of Totalitarianism*, and, while both arrive at a different notion of biopolitics,[4] the outcome of their respective perspectives builds on Arendtian theories: the notion of the camp as the inevitable logic of a biopolitical modernity for Agamben, and the significance of birth as rebirth in an affirmative biopolitics for Esposito, for example. Yet both authors find fault with Arendt as a biopolitical theorist. Agamben critiques the fact that Arendt fails to isolate a biopolitical perspective in her comprehensive work on totalitarianism and considers this a grave neglect in light of her other work, specifically in *The Human Condition*, in which she 'analyzed the process that brings *homo laborans* – and, with it, biological life as such – gradually to occupy the very centre of the political scene in modernity' (Agamben 1998: 3). Esposito, on the other hand, reproaches Arendt for having a 'blind spot' with regards to biopolitics: 'where there is an authentic politics, a space of meaning for the production of life cannot be opened; and where the materiality of life unfolds, something like political action cannot emerge' (Esposito 2008: 150). He asserts that it is indeed Arendt's lack of thinking the category of life through sufficiently which prevented her from a more accurate interpretation of life and politics. These are assertive claims and, coming from two highly prominent scholars of biopolitics, this criticism indeed raises the question: is an Arendtian account of biopolitics even thinkable?

Agamben draws considerably on Arendt's work for the concept of bare life in the figure of the refugee and the camp as the manifest logic of modernity, yet he is somewhat inequitable toward Arendt as a biopolitical thinker. As I have shown earlier, her assessment of totalitarianism was indeed more faceted than Agamben concedes. Furthermore, a political theorisation of an Arendtian biopolitics as the trajectory of the shift of life into politics, as traced above, requires a contextualisation of the work she offers on life politics in *The Human Condition* and *The Origins of Totalitarianism* with other texts she produced, to better understand the consequences for politics and the ethics of violence as a political tool in contemporary modernity. Where Agamben sees Foucault as completing that which Arendt seemingly has ignored, I suggest that Arendt indeed provides us with a basic and historic understanding of a trajectory of life as politics that, as the next chapter will demonstrate, converges in a many aspects with the biopolitical account Foucault delivers and on which Agamben builds his own theories.

Perhaps more pressingly at this point, however, is the issue of politics in Arendt's biopolitical account. As Esposito highlights in his criticism of Arendt, the issue of how the social realm and its preoccupation with life necessities constitutes simultaneously a sphere of politics remains somewhat elliptical. Arendt asserts that that which belongs to the private realm is by default disassociated from the political sphere. The perpetual cycle of biological needs and life necessities, and with it the labouring processes, for Arendt, exclude the possibility of politics proper because the homogeneous nature of biological life processes does not allow for plurality, which we recall as the primary condition for political action. How then, can we square the circle of this apparent contradiction in terms: bio*politics*? Arendt's work on what should or should not be a matter of politics has received much criticism and remains a controversial issue today.[5] While I acknowledge that the issue of what constitutes politics is complex in Arendt's work, particularly when viewed from a contemporary perspective, this is not the place to elaborate on the controversies or even delve deeper into whether she is indeed right or wrong with her strict categorisations. This is a task for some other time. What I intend to show in the remainder of the chapter instead is that there is indeed a way to understand Arendt's strict categorisation of biological life processes as labour in modernity as 'life-as-politics', or biopolitics.

Politics proper, for Arendt, is in close keeping with the Greek understanding of the term. It is an activity that takes place among a group of free (unencumbered by the demands of labour) equals acting and speaking in a public sphere. Political action comes to pass through action by unique individuals who present themselves in their particularity publicly, and through consensus building. The plural and contingent nature of different individuals coming together in a public realm means that political action is never entirely predictable in its final outcome and consequence, but always holds immanent the possibility for new, original beginnings, or *natality*. In short, contingency and uncertainty are immanent in politics proper for Arendt. This political activity is what distinguishes the purely biological form of existence from a more comprehensive idea of biological *and* political existence. It is, Arendt argues, in the capacity for political action, that humans distinguish themselves from both 'beast and god', and is as such the 'exclusive prerogative of man' (Arendt 1998: 22). Reliant on plurality, freedom and the existence of difference among humans in a public sphere, politics is – or should be – antithetical to the private realm of the household, which is chiefly occupied with managing life's necessities. Contra the public realm, this is a sphere exclusively dedicated to managing the basic needs of life, so as to then provide the necessary freedom for humans to engage in politics among their equals. Arendt explains: 'within the realm of the household, freedom did not exist, for the household head, its ruler, was considered to be free only in so far as he had the power to leave the household and enter the political realm where all were equals' (1998: 32). In Arendt's account, referring to the ancient Greek model of living together, the 'natural community' of the household was born out of necessity and it is biological necessity that is its sole focus. In the household, not plurality but the primacy of biological needs matters. Moreover, the household form, built on implicit hierarchical structures of inequality and role assignations, precluded the very possibility of providing a sphere in which equality is granted by its members to its members. In sum, where life processes dominate conditions of freedom, plurality and difference cannot be provided, and authentic political action becomes problematic. In its stead, the household format is scaled up to incorporate the wide-scale administration of life matters as politics. Arendt argues that in modernity 'we see the body of peoples and political communities in the image of a family whose everyday affairs have to be taken care of by a gigantic, nation-wide administration of housekeeping' (1998: 28). The *Sozialpolitik* of the social realm then functions like an enormous household administration, made manifest by the nation state, in modern politics. The nation state then represents a political form charged with matters of life processes and necessity, the very matters that are for Arendt excluded from political concerns. We begin to sense that a shift in the meaning of 'politics' is relevant here.

To better understand the political form of a biopolitically focused nation state I would like to suggest a term that Andrew Schaap employs in his analysis of the politics of need: an anti-political politics (Schaap 2010: 56–74). In his critique of Arendt's analysis of the labouring processes as non-political in modernity, he frames her argument in terms of an anti-political politics of need. While I differ in my assessment of Arendt's analysis of life politics from Schaap in his essay, the concept of an anti-political biopolitics perhaps best illuminates how the modern preoccupation with life processes can be understood as life-as-politics. While matters of life urgency and with it *animal laborans* excludes politics proper, it takes place within a formal political organisation represented in the social sphere in modernity, that is to say among humans, regulated by humans, managed by political governments. While the social subsumes the political, it simultaneously changes dramatically what politics was meant to represent in modernity. What we, in the present day, consider to be forms of politics, and specifically those forms of politics to which biopolitical analyses often refer (liberal, democratic, national and so on), specifically as they relate to contemporary issues of biopolitics, are, for Arendt, more akin to governmental administration and bureaucratic rule. Technically, then, we are faced with what amounts to a definitional problem in considering Arendt as having conceived of a bio*political* analysis, as, for her, the shift of life into the public (social) realm also meant an impediment of the faculty of action and a concomitant anti-politicisation. However, if we accept the dual notion that, in an Arentian sense, the biopolitics of modernity is inherently anti-political, the concept of biopolitics in Arendt's analysis is coherent with her perspectives of what belongs and does not belong to politics proper and is easier to follow. Chapter 3 builds on this differentiation and develops the consequences of an anti-political politics further.

Conclusion

This chapter has situated Arendt as a prescient thinker of biopolitical modernity. In her own project of understanding a complex modern reality, Arendt was primarily concerned with tracing the shifts that have produced a political reality which stands in troublesome ways under the sway of biological processes, informed and reproduced by a scientific-technological turn in politics. Arendt's analysis of life-as-politics can best be understood as a critique of liberal biopolitics in modernity. Her analysis and exposition of the modern shift of life into the political centre, through the manifestation of the social realm, presents a thorough understanding of the main biopolitical facets as they manifest within contemporary discourses on the topic. While Arendt focuses neither on the role of the sovereign nor on distinct institutions associated with biopolitical mechanisms, she presents in her work a thorough consideration of the category 'life' as both the basis and the target of political forms in modernity. This allows us to more clearly identify how contemporary techno-biological subjectivities are shaped and conditioned by a biopolitically informed rationale. It is this biological-technological rationale that Arendt decidedly put her finger on, which provides the cartography for contemporary practices of political violence and their ethical framing. Both the *zoe*fication of politics and the politicisation and technologisation of biological life, or *zoe*, are relevant aspects in considering biopolitical subjectivities and thus have decisive consequences. Both aspects of Arendt's strands of biopolitics then harbour significant implications for the modern conflation of violence with politics, and the ethical justifications with which they are narrated.

Arendt convincingly delineates the emergence of the human as determined by labour to sustain life processes – the life processes of the biological body and the life processes of the sociopolitical body. Here, economic cycles of production and consumption mirror the life processes of biological existence and place large swathes of life under the mandate of political management. In this, the human, a labouring entity, is functionally embedded. Following on from this, with Arendt, we can also identify the trajectory upon which humans came to be viewed, and view themselves as an analysable and mathematisable entity, with an ever greater reliance on measuring instruments, algorithmic calculations and the authority of 'code, data and statistics' (Ansorge 2012: 2). This opens the door for a more thorough understanding of the co-constitutive nature of biology and technology and its consequences for political violence today.

Having set the stage with an excavation of biopolitical thought in Arendt's work, I will make this my focus in later chapters. In this I will engage in greater detail with the implications this specific biopolitical rationale has for perspectives of and possibilities for politics and how this relates to the use of violence. Where some scholars see an impasse in reconciling biology and politics in an Arendtian account, I suggest that it is indeed possible to establish a coherent account of biopolitical modernity in Arendt when we consider crucial distinctions in Arendt's conception of politics with those of a contemporary perspective of politics primarily as professional management and administration. In understanding Arendt's analysis of biopolitics as an essentially anti-political form of politics, a form of professionalism, in modernity, it is possible to see beyond this alleged impasse. This is discussed further in Chapter 3. Drawing on Arendt for a comprehensive account of biopolitics demands, however, an engagement with the most eminent source for biopolitical critiques: Michel Foucault. This follows next.

Notes

1 See for example Campbell and Sitze 2013a; Agamben 1998; Braun 2007; Duarte 2006; Esposito 2008; Fassin 2009; Hayden 2009; Vatter 2009 among others.
2 The second edition of *The Origins of Totalitarianism* was published in 1958, the same year *The Human Condition* was published. This second edition contains a few changes Arendt made as more material had been made available in the years between 1951 and 1958. Most importantly, her essay 'Ideology and Terror: A Novel Form of Government', first published in 1953 in the *Review of Politics*, was added to this second version as it contained theoretical material relating to her analysis of totalitarianism, terror and domination. The timeline of this important addition suggests that Arendt's theories on the human condition have informed elements of this final chapter and vice versa.

3 The social is most often understood as the realm in which economic concerns gain political relevance and it is precisely her dismay at economic concerns becoming politically relevant which has caused much debate among Arendt scholars. Particularly Arendt's conception of the social as a realm in which biological concerns dominate has received much criticism and, even until recently, scholars have been concerned with trying to establish definitively what Arendt meant with this sphere. See Fenichel-Pitkin (1998) for an extensive account as well as Owens (2012) for an analysis of the social in the context of security. The social is indeed not entirely well defined in Arendt's work, part of this being perhaps an issue of inconsistent semantics on her part. However, the social as a concept in the discussion of forms of politics in modernity sheds an important light on the quasi-spatial aspects in which human life unfolds. Another key criticism of Arendt's work is her critique of Marxism which runs through *The Human Condition* and other works where some scholars have held that she based her critique on an outmoded interpretation of Marx or misunderstood Marx altogether (see for example Bernstein 2006; Jay 2006).

4 Agamben frames a predominantly negative and critical account of biopolitics in his scholarship, while Esposito seeks to rescue an affirmative potentiality in biopolitics, specifically through his account of natality (Agamben, *Homo Sacer*; Esposito, *Bios*), and both draw on Arendt to do so.

5 See for example Calhoun and McGowan (1997); Schaap (2010); Bernstein (2006)

2

Biopolitical technologies in Arendt and Foucault

As a category of revolutionary thought, the notion of historical necessity had more to recommend itself than the mere spectacle of the French Revolution ... Behind the appearances was a reality, and this reality was biological and not historical, though it appeared now perhaps for the first time in the full light of history.

Hannah Arendt, *On Revolution* (2006a: 49)

[W]hat occurred in the eighteenth century in some Western countries ... was a different phenomenon having perhaps a wider impact than the new morality; this was nothing less than the entry of life into history.

Michel Foucault, *The Will to Knowledge* (1998: 141)

While the focus of this project centres on the biopolitical ideas in Arendt's work, it would be an oversight not to engage with Foucault, as the proverbial father of a theory of biopolitics, in order to delineate more clearly where Arendt's work on life politics in modernity might provide an additional insight into how biopolitical structures inform our understanding and acceptance of violence today. Arendt and Foucault appear alongside each other in a wide range of contemporary scholarship,[1] and the two thinkers converge in many aspects of their respective analyses of the challenges of modernity: both emphasise the distinctly modern shift of biological life into the centre of public concern, which provides the political underpinning of a monolithic collective, and both use a historical trajectory to approach their respective analysis of how modern structures have emerged.[2] With that in mind, this chapter engages with the key features in which the Arendtian and the Foucauldian models of biopolitics show parallels, and in which aspects they differ. The chapter looks specifically at some key elements that characterise and determine the biopolitical perspectives of both. Finally, I then identify some key contributions the Arendtian account of biopolitics can offer for an analysis of political violence in contemporary technological modernity. But before delving more deeply into the symmetries and asymmetries of the respective accounts of life politics in Arendt and Foucault, it is helpful to briefly establish the cornerstones of Foucault's account of biopolitics in modernity.

Where the term biopolitics appears in contemporary discourses in of politics, the name Michel Foucault is typically associated with it. Foucault was one of the eminent analysts of the subject 'life' in modernity and has essentially provided the foundation for much of the biopolitical examinations of power structures and the production of knowledge in modernity. His analytical framework of power, constituted as bio-power in Western modernity, has been much referred to and is widely used by politics scholars (most enthusiastically perhaps by critical theorists) in explaining and understanding national and international political and social relations. While he was not the first to use the term biopolitics, his theories are dominant in the biopolitics discourses in political theory today. To understand why Foucault has provided such a novel and persuasive lens with which to analyse the role and function of life-as-politics in a complex contemporary modernity, it is helpful to briefly examine Foucault's aim of analysis in which he seeks to examine the 'modes of objectification that transform human beings into subjects' (Foucault 2002b: 326) and the concomitant trajectory of power, beginning with the docility of bodies, as instituted through disciplinary powers in the late eighteenth century.

Disciplinary tactics, biopolitical strategies

In his detailed account of the formation of malleable bodies in Discipline and Punish, Foucault lays the groundwork for a technology of power and its political investment in which the physical body becomes subject to punctilious methods of control, instituted through temporal and spatial forms of restriction (Foucault 1991: 135). The natural body becomes the target, not en masse as an indivisible entity but in its various distinct mechanisms and functions, thus making the body 'analysable' and subsequently 'manipulable'. In order to wield this power effectively on the body, the physical body needs to be rendered in numerical and mathematical terms, so as to apply measures and evaluation for a progressive development of the body.[3] Through continuous and meticulous exercise, temporal measurement, spatial restrictions and the establishment of the notion of rank, the natural body is thus shaped into an instrument for efficiency, productivity and achievement, it becomes insertable and replaceable, in short, a commodity. Foucault importantly notes that this particular 'control of the operations of the body' differs from slavery in that it is not an appropriation of the body which is achieved through violent coercion, nor is it comparable to the regulations of 'service', with a 'non-analytical, unlimited relation of domination' (1991: 136–7). Rather, these techniques of 'docility-utility' exerted upon the human body are 'directed not only at the growth of skills, nor at the intensification of its subjection, but at the formation of a relation that in the mechanism itself makes it more obedient as it becomes more useful and conversely' (1991: 137). The power wielded here is for Foucault in essence a political matter, as the disciplinary practices became 'formulas of control' (1991: 139). The body thus is rendered subject to the politics of docility, and made into an object of utility in this political context. It is a political objectification of the physical subject through these anatomo-political technologies, from which 'man of modern humanism was born' (1991: 141). This politics of the anatomy, however, is for Foucault only one pole around which 'the organization of power over life' (1991: 138) was installed. The second pillar was the growing mechanics of power over the population as a wholesale entity, a political power that Foucault first named 'biopolitics' in volume 1 of The History of Sexuality.

In conceiving of biopolitics, Foucault shifts away from the disciplining of the individual docile body through codified practices of control that divide relations of space and time, and towards a politics that controls the health, maintenance, organisation and management of populations. Biopolitics is thus a technology of power that exerts its reign not solely on 'anatomo-politics' but on the politics of the human race (2004: 240). It is not the discipline, surveillance and control of the individual physical body that is of concern but the health and safety of an entire population, its birth rates, mortality rates and longevity (1991: 243). The preservation and security of the human race becomes the basis for a new type of discipline for the regulation, organisation and management of a body of multiples. As such, the population becomes an economic and political resource for the perpetuation of production and consumption; an asset, the life of which must be preserved, increased and extended. Control over the life of populations is thus exercised through the institution of measures, statistics, forecasts, categorisations and rankings (1991: 246).

As in the Arendtian account, a *mathematisation* of life and life processes fundamentally underpins the Foucauldian analysis, although he does not explicitly establish this in his work. These practices employ the systematic collection of knowledge and 'hard facts', so as to make the disorganised, messy and unruly mass of people organised and manageable (1991: 54). As a result, it is not individuals who, collectively, shape and define a population, but rather the population that defines the individual through naturalised modalities and regulations placed upon the individual. Individuals then, in turn, are shaped and defined by the practices and discourses related to the promotion of health and management of populations, act in accordance with the practices and discourses, and thus come to govern themselves in a perpetuating and reflexive manner. In his essay 'Governmentality' Foucault states the matter of self-government in such systems of biopolitical modalities as follows:

> In contrast to sovereignty, government has as its purpose not the act of government itself but the welfare of the population, the improvement of its condition, the increase of its wealth, longevity, health, and so on; and the means the government uses to attain these ends are themselves all, in some sense, immanent to the population; it is the population itself on which government will act either directly through large-scale campaigns, or indirectly, through techniques that will make possible, without the full awareness of the people, the stimulation of birth rates, the directing of the flow of populations into certain regions or activities and so on. (2002a: 217)

Biopolitics differs from the discipline of the sovereign in the practices it employs, as it does not seek to alter the individual as a physical body, but rather impose a generality on to the individual as an element of a population. However, as Foucault points out, the practices and institutions of biopolitics do not entirely replace the disciplines that exist for the control of the individual malleable body, rather they exist on a different, supplementary level, building a more comprehensive network of power that reaches to the far levels of societal structures (2004: 270). This creates a matrix that serves to control the population, and thus the individual as an element of a population in a multitude of aspects in their lives (Jabri 2007: 45–62).

The modern humanity Foucault traces in his work gives rise to a central question for him, and in his lecture series published in English under the title *Society Must Be Defended* he asks: 'how can the power of death, the function of death, be exercised in a political system centred upon biopower?' (2004: 254). In a system that hails the preservation, proliferation and security of the comprehensive human race, Foucault introduces a variant of the concept of racism as the binary counterpart to biopolitics, the politics of death, of letting die that which is to be located as outside of and a danger to the human race. As such, racism has two crucial dimensions for Foucault: firstly, to separate 'the field of the biological that power controls' in order to disconnect various elements of a population into sub-populations, thereby creating a binary structure (2004: 255). Secondly, the institution of racism serves as a validation of the prevalence of one (good) race over another (bad) race. The relationship becomes biological, 'the death of the bad race, of the inferior race … is something that will make life in general healthier: healthier and purer' (2004: 255).

While a Foucauldian theory of biopolitics in its decentralising structure essentially undermines the role of sovereign power in the traditional sense – the legitimised power to 'take a life or let live' (2004: 241) – there remains the requirement for a decision, whether that be the decision to engage in an act of war or any other decision that affects the very biopolitical remits of a population. In this, Foucault points to Nazi Germany where a nation state has asserted its sovereignty through the employment of biopolitical tactics that eventually, if not necessarily, enable genocidal practices. In Foucault's words: 'The Nazi State makes the field of the life it manages, protects, guarantees and cultivates in biological terms absolutely coextensive with the right to kill anyone, meaning not only other people but also its own people (2004: 260). As Foucault explains, in Nazi Germany the totalitarian administration utilised biopolitical modalities to establish its own biopower, based on the exclusion and subsequent elimination of that which was deemed 'inhuman' to allow the human race to prosper. Technological means of communication and media were employed by the state to propagate the Aryan ideology and to depict specific non-Aryan ethnicities as vermin, as animalistic, as inhuman, so as to facilitate the normative exclusion and the subsequent state of exception (*Ausnahmezustand*) in which legal restrictions to act upon the established discourse were suspended.

In this construct, the imposition of exclusionary practices, however, takes place no longer at the hands of a single sovereign but through multiple networks of both disciplinary and biopolitical institutions. As Foucault recognises, disciplinary power and biopower thus perform a supplementary role to one another, in a web of power that reaches into the very interstices of societal levels (2004: 250). It can only be through the creation of the docile, manageable body that the preservation of a distinct and manageable population is enabled, and normalising practices and discourses for the preservation of the population can be instituted. Perhaps comparable to Arendt's *animal laborans*, the Foucauldian subject is conditioned to become 'submerged in the over-all life function of the species' (Arendt 1998: 322), the disciplined individual is accordingly shaped into a 'fighting machine, a working machine, or indeed a consumer machine' (Jabri 2007: 57), to consume for the promotion of a population, to work for the proliferation of a population or to fight wars for the protection of a population. In short, the individual is rendered an object of utility for that which sustains and shapes that very object as subject. It is through the institution of political economy in conjunction with immanent disciplinary technologies that Foucault's power *dispositifs* come to establish a range of relations between the power structures that reach through the social body and materialise within the minutiae of societal relations. The synergetic constitution of disciplinary tactics and biopolitical strategies hence frames the procedure in which the contingency of human life is

shaped into productive and consumptive life, while producing the biopolitical *dispositifs* that enable a population to wage war in pursuit of its
preservation.

Relevant to the investigations in this book, there are two aspects in
Foucault's analysis that are thus critical to his biopolitical perspective: the
body as a subject to, and of, political power structures, and the perpetuation of the practices and norms that shape the individual and collective
body through institutions. Both work on, and through, the subject. This
raises the question of 'what is the subject in Foucault', a question that has
widely been problematised. Amy Allen takes up this issue in response to
what a wide range of scholars see as Foucault's position as anti-subjectivist
(2000: 115). This position posits that, in Foucault's analysis, the subject is
merely a husk for the power inscriptions of anatomo-political and biopolitical technologies. As such, the Foucaultian subject is entirely constituted
through the discourses and practices that work upon the body and the
population – in short, the anti-subjectivist position holds that Foucault
does away entirely with the subject as such – it becomes an object only
(2000: 116). Allen argues against this position in that she points toward a
nuance in Foucault's work which helps understand that, far from declaring
the subject dead in his analysis, Foucault must be read as wishing to investigate what constitutes and explains the subject in modernity, rather than
how the subject itself constitutes subjectivity (2000: 121–3). In Allen's
words: 'Foucault indicates ... that the goal of his analysis is to ... explain
the subject as constituted through discourse, to shift the subject from the
explanans to the *explanandum*' (2000: 123). Thus, in order to explain the
subject, Foucault's inquiry would have to move away from an understanding of the subject as constitutive. In other words, the 'inside', the transcendental of the subject, is deliberately left to one side in order to explain the
'outside' of the subject (Deleuze 2006).

Allen's position is helpful in stressing Foucault's specific vantage point
of inquiry and confirms his explicit aim of analysis to be the '*modes*
of objectification that transform human beings into subjects' (Foucault
2002b: 327, emphasis added) but it ultimately fails to elucidate subjectivity,
and with it agency, in the Foucauldian account. There are others (see for
example Bernstein 1992; Patton 1998; Shinko 2008) who seek to rescue the
subject from its Foucauldian neglect, by turning toward Foucault's later
writings on ethics and the care of self. And, indeed, in these writings the
subject appears as a much more 'active' subject than previously in which
consciousness and reflection are present, and with them a certain possibility for agency (Foucault 2002b: 291). However, as Deleuze has effectively
explained, the essentially hollow 'inside' of the subject is constituted only

and always by the doubling folds of the outside (Deleuze 2006: 78–101). Foucault is very clear about this himself: he rejects a theory of the subject as a lens for his approach and states instead that the subject is not a substance at all, but rather should be understood as a form (2002b: 290). Based on the contingency of these forms there are, then, multiple selves that depend on the constituent context. For Foucault, there are two meanings of 'subject': 'subject to someone else by control and dependence, and tied to his own identity by a conscience or self-knowledge' (2002b: 331). The existence of a conscience would suggest the existence of subjectivity, however, if we follow Foucault's assessment of the 'inside' of the subject to be shaped and informed by the truth games and relations of power of the outside, then this consciousness is only ever a recognition of that which has been imprinted as constitutive of a form – which is essentially contingent.

This leaves us with an unsatisfactory account of the agency of the subject to resist relations of power and, specifically in in light of biopolitical technologies of power, with an impasse for resistance. The perpetual power dynamics Foucault describes, which run through the interstices of society as the matrix of power, permit no exit from this matrix. Rather, the Foucauldian dynamics of power reinforce existing structures along which biopower operates. Rupture of these biopolitical *dispositifs* is not a distinct possibility in the Foucauldian account, and subjects, conditioned and shaped through anatomo-political and biopolitical modalities as they are, have an already always conditioned (and thus limited) scope for action against the structures that shape them. An account of political action as a creative act, or a productive force is missing in Foucault. His analytical concern is to look at how power is produced and circulates, and how the political subject is constituted through this. Arendt, in contrast, is concerned with the (im)possibility for politics in the modern human condition. In this, Arendt's account of life politics in modernity, in contrast, may have a greater potential for such a caesura through specific concepts shaping the human condition, such as *natality* (Blencowe 2010; Lupton 2012; Vatter 2009) or a perspective of temporality in Arendt's biopolitics (Braun 2007).

This is just one fundamental difference in the broader perspectives on life-as-politics in Arendt and Foucault. In the following I will highlight some substantial and, for the purpose of this book, significant overlaps in both analyses, then I will assess some of their ontological premises relating to these accounts in which both convergences and contrasts can be identified. Finally, I will be engaging more closely with the distinct differences in their respective insights into life-politics in modernity, so that we can better understand where Arendt offers an additional insight for a contemporary understanding of biopolitics.

Symmetries in Arendtian and Foucauldian biopolitics

I have noted earlier that scholars of political theory and international relations alike have read the works of Arendt and Foucault as both complementary and contrasting.[4] This is not surprising, given that both offer among the most piercing and useful critiques of politics in the modern era, and both seem to have recognised and articulated early the challenges that face us in the contemporary political world. Both Arendt and Foucault consider 'the political' in the modern era as one determined predominantly by social, economic and technological structures, and both, at different junctions of their respective work, express their bewilderment at modernity's dark puzzle: how can an era that claims to value life above everything else as a public concern simultaneously be one that has the greatest potential for genocide and mass atrocity (Dolan 2005: 370; Isaac 1996; Braun 2007)?

Comparing two scholars as different in their approach and perspectives as Arendt and Foucault is, however, fraught with the risk not only of comparing apples with oranges but also of perhaps drawing connections where there are none, and of ultimately missing the essence of either account. It must be a goal of any analysis that reads these two together to recognise that Arendt and Foucault each had a different focus for their analyses, despite the apparent fact that both chose to analyse the challenges of modernity from a historical-philosophical stance. Where Foucault places emphasis on the genealogy and properties of power and knowledge for his analysis of the historical processes by which biopolitics materialised (Cooter and Stein 2010: 110), Arendt is chiefly concerned with an evaluation of the human condition and the possibility for politics proper in modernity when she traces her assessment of life-politics in the modern era. It is thus perhaps more accurate to speak of symmetries, or parallels, in discussing the similarities in the two scholars' accounts of modern life and politics, so that we are able to retain and recognise the value of each analysis to the wider field of biopolitics. A survey of their respective analyses of modern life-politics flags up a number of obvious parallels: both highlight the emergence of the category 'life' as a resource for cycles of production and consumption that engender a modern political economy of life; both suggest that the shift of life into the political centre results in a new paradigm of security; both establish clearly that, in a life-centric political modernity, the individual is amalgamated by the generalised body of the population as a whole; both discuss the management of life as the chief political activity and the prevalence of administration-as-government, largely subsuming human social and political activities; both suggest that, in such a context, life is characterised primarily by utility and economic value; and both are critical of the institutionalisation of science in the production of knowledge. A thorough examination of all parallels and overlaps would make an interesting topic for a different book, but is not the aim here. Instead, I will highlight two key aspects – historical context and mechanisms – which underwrite the modern account of biopolitics in both scholars' work, and form an important backdrop to their respective analyses.

The historical-philosophical approach both scholars favour in their analyses of biopolitics identify a specific turning point in modern political history. Both Arendt and Foucault identify the late eighteenth century, and specifically the French Revolution, as pivotal in the emergence of life necessities as the central object of politics (Edwards 1999: 7). For Foucault, it was with the French Revolution that a modality of power over biological necessities – biopower – came to be codified in law. This codification institutionalised 'continuous regulatory and corrective mechanisms' whereby juridical structures became the normalising power for a range of regulatory apparatuses, making 'an essentially normalising power acceptable' (Foucault 1998: 144). Like Arendt, Foucault clearly points to this specific historical context in which a new and increasingly self-referential social perspective emerged:

> One no longer aspired toward the coming of the emperor of the poor, or the kingdom of the latter days, or even the restoration of our ancestral rights; what was demanded and what served as an object was life understood as the basic needs ... It was life more than law that became the issue of political struggles, even if the latter were formulated through affirmations concerning rights. The 'right' to life, to one's body, to health, to happiness, to the satisfaction of needs. (1998: 145)

In this shift, biology appeared as an object of politics, which then came to be regularised and normalised through law as a normalising institution. It is against this background that the politicisation of the body, specifically 'sex as a political issue', becomes intelligible for Foucault (Ibid.). Where Foucault centres his diagnosis of the juridical-political regulation of the body and populations, Arendt identifies the same historical event as the moment when political theory became saturated with ideas of biological necessity as both a political claim and an organising principle for a common political body. In this moment the question of political freedom became one of biological necessity – the social question (Arendt 2006a: 49–53).

In both accounts, it is essentially the manifestation of biological needs and matters of life necessities, emerging through the French Revolution, that engendered a biopolitical dimension, gave rise to the possibility for institutions of racism and essentially bred a dangerous ideology of homogeneity, the perils of which Arendt highlights at length throughout her writings. For both thinkers, the French Revolution, as a pivotal event, provides the parameters for modern biopolitics. For Foucault the French Revolution represents the historical moment when the normalisation of anatomo-political and biopolitical power structures were codified in an unprecedented regression of the juridical (1998: 144). For Arendt, the French Revolution epitomises the historical moment when the plurality of political humans began to be in peril, as the uniform clamour to have basic biological needs met at a political level became the lowest common denominator of highest significance. Here, Arendt takes Marx to task and critiques him as a catalyst for

> the politically most pernicious doctrine of the modern age, namely that life is the highest good, and that the life process of society is the very centre of human endeavour. Thus the role of revolution was no longer to liberate men from the oppression of their fellow men, let alone to found freedom, but to liberate the life process of society from the fetters of scarcity ... Not freedom, but abundance became now the aim of revolution. (Arendt 2006a: 54)

It is from here that Arendt is able to analyse the precarious relationship between necessity and violence. Chapter 4 explores this tension further.

A second relevant symmetry in what Arendt and Foucault identify as central characteristics of a biopolitical modernity includes the observation that, in order for life to become central to politics, life and its processes needed to be rendered analysable, predictable and thus controllable. In other words, both thinkers place great significance on the scientifically transparent human for the realisation and perpetuation of life politics. The development and prevalence of scientific inquiries into the human body, so that it could be disciplined and regularised (Foucault 2004: 253), and into the human mind so that it can be 'tested and measured like horsepower' (Arendt 1998: 284), serve for both as a decisive condition for life-centric politics to be enabled and sustained. In both accounts, the mathematisation of life and its functions renders humans as statistical objects and subjects for the normalising processes of a biopolitical society. For Foucault, statistics, measures and forecasts serve as essential mechanisms to intervene into the body of the population for the promotion of life: birth rates must grow, mortality rates need to be reduced, life expectancy should be increased and the inherently aleatory and risky condition of life is to be stabilised and secured by statistically inferring relevant risk factors (2004: 246). Although Foucault does not make it explicitly clear, this focus on statistical rates and measures that determine what is normal and abnormal in life processes has become possible only with a specifically modern scientific approach to human nature. By rendering human biology in scientifically observable and measurable terms, we can become subjects to docility at the individual level, and regulation at the population level. While Foucault clearly identifies the scientific-technological biopolitical mechanisms that serve to 'achieve overall equilibration and regularity ... of man-as-species', he does not engage with the historical-philosophical turn that enabled this (Ibid.). He is thus not immediately helpful in understanding the modern human subjectivity that helped produce the possibility for such mechanisms.

Here Arendt's account is more elaborate than Foucault's, as she traces for us the development of this self-referential human-centric mathematisation of life, as the previous chapter has shown. In both accounts, however, the reliance on statistics, as the master category for utility, progress and processes, is norm-producing. Didier Bigo identifies the consequences of the Foucauldian ascendancy of statistics where '[p]opulation is not people. Population is a statistical category' (Bigo 2011: 99). The statistical capture and rendering of the population has two implications: it first furnishes the numerical basis for what is normal. And it further establishes that which is outside the normal margins – the abnormal – as a risk to the population. The category of the abnormal is then classified as risk to all that which is statistically normal, which in turn requires that it be politically secured. Bigo explains the mathematical logic of normalisation:

> Abnormalisation is derived from constituting statistical regularity and classifying procedures which distribute events into particular categories labeled by bureaucracies of the state. Normalisation is not carried out through some principle of division but through statistical distribution. (2011: 105)

Bigo's Foucauldian analysis of the normalisation process mirrors Arendt's assessment of the consequences of statistics as a dominant political measure in modern society. For Arendt, the prevalence of statistics as a key feature of the social realm in modernity presents a nearly inescapable danger of rendering society homogenous, simultaneously establishing that which is abnormal to the statistically determined homogeneity of society (1998: 42–4). In her analysis she distinguishes between acting and behaving. The former relates to the political realm, the latter corresponds to the social realm. The former cannot be measured, as it is inherently contingent on context, the latter can be scientifically captured. This is where statistics become most relevant for Arendt, as a mechanism for conformism. Where the population is understood as a resource for the political economy, and where political economy is understood to function in the image of a scientific process, the population can be captured numerically only in its everyday behaviour, not action. What is captured then as normal is already always determined by conformity. And the greater the number of conformity data points, the greater their validity, the most substantial the uniformity of behaviour and with it the normalisation of a specific behaviour. This in turn then produces what she calls a dangerous fiction of a uniform and harmonious society. She notes: '[s]tatistical uniformity is by no means a harmless scientific ideal; it is the no longer secret political ideal of a society which, entirely submerged in the routine of everyday living, is at peace with the scientific outlook inherent in its very existence' (1998: 43).

With statistics as the master guide for correct and regulated behaviour, non-behaviour, non-uniformity becomes increasingly less likely to be tolerated. Dichotomies of acceptable and unacceptable, safe and dangerous emerge, based on a scientific norm from which then springs the possibility of 'correct prediction' (1998: 43). Although a critique of the technocratic understanding of society in modernity and a strong scepticism of the institutionalisation of science as a producer of knowledge are present in both Arendt's and Foucault's works, Arendt is much more explicit and distrustful of the general advent of the sciences as a new means for truth-finding in modernity than Foucault. However, both the Foucaultian and the Arendtian accounts rely on the prevalence of algebra, mathematics and statistics as a central modality that provides the basis for the practices that determine and shape society in the modern era.

Both the operative mechanism of statistics and the historical rupture in eighteenth-century politics make for an interesting contextualisation of Arendt and Foucault's work. To better understand what informs their respective accounts and analyses, a brief engagement with their respective perspectives on the human, on history and on power is useful.

Essence, telos, power

When reading Arendt against Foucault, or vice versa, an interesting consonance on some of the ontological premises of each account becomes apparent. These relate to their perspectives on human nature, on history and on the quality of power. And as much as there are congruencies, an examination of their respective position also reveals the nuances in each account. Firstly, both Arendt and Foucault reject, in principle, an essentialist view of the human, emphasising instead that human beings are fundamentally shaped and influenced by their social environment (Gordon 2002; Braun 2007). For neither Arendt nor Foucault is there an immanent ontological core to human nature that is independent from the social relations and conditioning factors of the environment they are situated within. This intrinsic condition of the human as constituted and shaped by social and political forces is perhaps somewhat more clearly a stipulation for Foucault than Arendt. Foucault distinctly rejects the notion that there is an essence to the human, which could, or should, be liberated. He notes:

> If [liberation] is not treated with precaution and within certain limits, one runs the risk of falling back on the idea that there exists a human nature or base that, as a consequence of certain historical, economic and social processes, has been concealed, alienated or imprisoned in and by mechanisms of repression. (Foucault 2000: 282)

For Foucault, the idea of liberation is tricky when it aims at revealing the 'true' human, as, for Foucault, such an entity does not exist. Rather, humans are always constituted through regimes of power that manifest through the individual body and the body politic more broadly. The non-essentialist account in Arendt's work is more complex as it is not as clearly inscribed into the physical body in her work. Arendt remains, in fact, conspicuously silent on the human body as a site of inscription. Furthermore, there could be considered an element of essence in Arendt's work: each human being is born with the intrinsic capacity for a new beginning. In this sense, the immanent uniqueness within each beginning may present an essence of the human as such (Canovan 1995: 70). It is in fact with this uniqueness that each individual holds that new beginnings in turn are created in Arendt's account. However, it is also precisely here that any essentialising moment ends for Arendt, as the uniqueness of each individual is further shaped and determined through a material ecology and through the various contingencies that interactions with others carry within them. The human, as such, remains in flux. Beginning, then, may perhaps more accurately be considered the essence of the human condition, rather than human beings *per se*. Canovan clarifies this distinction as she notes: 'it is of the essence of the human condition that the persons who inhabit the world are continually changing' (Canovan 1995: 130). For Arendt, it is on this basis that the plurality of human beings can be maintained, and the capacity for action ensured.

This constitutes also a significant difference between Arendt and Foucault: the non-essential human nature, the perpetual dynamics of contingency that they, humans, are subject to, is simultaneously the essence of the condition of action as politics for Arendt. It is in the capacity for new beginnings (natality), through action, that we can potentially resist and revolt against existing power structures in a biopolitical context. And even though the capacity for action is in grave peril in an Arendtian biopolitical modernity, there is always the hint of a possibility for a new beginning towards something hitherto unknown. As I have pointed out earlier, this dimension is limited at best in Foucault. The subject is always objectified in biopolitical structures and this objectification constitutes the subject. While there is resistance, this resistance always relates only to these existing power relations. It is this non-essential quality of humans that enable disciplinary and biopolitical power structures to form and shape humans and thus render them perpetually conditioned under the mandate of power-exerting structures.

Similar to the non-essentialism evident in Foucault's and Arendt's biopolitical perspectives of the modern era, they both also strongly reject any teleological understanding of history (or nature) (Gordon 2002; Braun 2007; Dillon 2011). Foucault is highly sceptical of any notion that would dissolve the dynamic and perpetual events of particularity into a greater metanarrative which eventually reduces all that unfolds among people to a self-referential totality. The endlessly continuing matrix of power structures renders the fulfilment of such a deliberate trajectory of history implausible, as Foucault prioritises the 'genealogist's "single drama" over progressive "evolutionary" paradigms of the Enlightenment and dialectical materialism' (Roulier 1997). Similarly, Arendt fervently rejects a teleological understanding of history or nature as a linear process. For her, any understanding of history as a history of movement, of history as a 'constant flow, concerned with the development of mankind and continuous with the evolution of nature' was a modern and highly anthropocentric invention (Canovan 1995: 75; Villa 1999: 5).

Like Foucault, Arendt did not favour the Hegelian notion of history as a gigantic process, subsuming individual stories and events into one larger realm of meaning. She rejected this not only on the grounds that such a perspective has proved to be dangerous as an instrument in the emergence of totalitarian forms of politics but also as the idea of a teleological fulfilment of history as a process runs counter to action and contingency, both crucial Arendtian theoretical categories (Canovan 1995: 76–7). The condition of natality means that history cannot and ought not to have a telos, let alone one that is made and determined by humans, for humans. For Arendt, the rejection of history as a process is perhaps much more fundamental to her critique than it is for Foucault. Recall that for Arendt it was the modern understanding of history as a process, and its strong relation to the perception of nature as a process towards progress, which provided the dangerous grounds for the murderousness of totalitarianism. This connection is absent in Foucault.

The third fundamental aspect where the two scholars intersect is in their understanding of power as opposite, if not contrary, to force or strength (Arendt 1970: 43; Foucault 2004: 29). However, here the intersections are perhaps fewer than the ultimate differences between the two accounts of power. The distinction both make between power and force, domination or strength is, however, vital. Power, in Arendt and Foucault is a property that functions and is actualised. In other words, power, for Foucault and Arendt is not an instrument of force individuals are subjected to, rather, power is generated among people (Arendt) or circulates perpetually (Foucault). Power is boundless in both accounts, its only limits are other people (Arendt 1998: 201; Foucault 2004: 29) and, for both Arendt and Foucault, power is entirely relational and takes place between subjects. In other words, both maintain that power can be neither possessed nor stored. Furthermore, for power to have functionality, the subjects involved in the power relationship must first be free in both accounts. As we have seen earlier in Foucault, power is brought to bear on the freedom of a subject, requiring docility for complicity in the exertion of power. Similarly, in Arendt's account, freedom plays a vital role in facilitating the coming into being of power. Being subjected to the force and domination of a master does not render the master powerful nor is the subject one that is free. Freedom, like power, for Arendt can take place only in plurality. In order to come together and act in concert, subjects generating moments and circumstances of power must have gained freedom within the context of a plurality.

Although this is an important premise for their respective perspective of biopolitics this is, however, the limit of the extent to which the two thinkers overlap in their theories of power. There are finer nuances between the two theories of power that differ substantially in each account. In Foucault's analysis, power is constitutive and relational among multiple elements. For Foucault, identities are shaped through power and power is articulated through the subject (Gordon 2002). Dillon and Neal place emphasis on the multi-layered complexity of power in Foucault:

> Power is ... palimpsestuous. It is inscribed on us; we also inscribe it on ourselves, through many institutionalized writing practices. The outcome is a mobile, mutable and complex manifold in which different formations of power are continuously at play in different ways throughout different aspects and formulations of life. (Dillon and Neal 2011: 13)

The positive production of power in Foucault's analysis requires that the individual subject displays a certain level of complicity in the ordering and controlling practices of disciplinary power. Michael Hardt and Antonio Negri effectively highlight the key contribution Foucault makes here: 'Foucault was insistent on the fact, and this was the brilliant core of his analysis, that the exercise of discipline is absolutely immanent to the subjectivities under its command' (Hardt and Negri 2001: 326). And indeed what stands at the very core of Foucault's understanding of the workings of power is the differentiation that power is not exercised through coercive measures, by sheer violence, but that it is wielded by shaping a free subject's behaviour through certain technologies. Foucault illustrates this point in one of his lectures in 1979:

> A man who is chained up and beaten is subjected to force being exerted upon him, not power. But if he can be induced to speak when his ultimate recourse could have been to hold his tongue, preferring death, then he has been caused to behave in a certain way. His freedom has been subjected to power. (Foucault 2002b: 324)

Power, for Foucault, is passed through individuals and articulated at a specific site only for as long as it is active within these particular power dynamics. Power perpetually continues to circulate. For Foucault, power has functional properties and works through networks. In his words: '[power] is never localized here or there, it is never in the hands of some, and it is never appropriated the way wealth or a commodity can be appropriated' (Foucault 2004: 29).

In distinction to Foucault's theory, power for Arendt is generated among people and is fully tied to the subject. The locus of the origin of power is the coming together of people in acting on a shared interest. Power can thus exist only for as long as this collectivity of power-generating subjects remains an active collective. A disbandment of the collective lets power vanish. For Foucault, however, even though power is exerted always temporarily only at a specific site, power does not vanish or disintegrate, rather, it shifts locale. Like energy, the Foucauldian notion of power is only ever transferred and does never truly dissolve. Staying within a terminology of physics, the Arendtian concept of power, in contrast, is more closely related to a process of friction by which power is generated, but the level of this generation of power cannot be institutionalised permanently. In other words, power can only ever be 'actualized', but never fully 'materialized' (Arendt 1998: 200). It always relies on the plurality of people. This means that power, for Arendt, can be generated at different locales and be maintained wherever people come together through shared interests, including in the resistance to other locales of power. For Foucault resistance to power is problematic, as it always shapes the subject in a continually oscillating transference of power that forms the subject's interests. As Neve Gordon explains, '[f]rom within Arendt's imaginary one would have no difficulty explaining how a group of students can decide to resist the powers that be, yet a Foucauldian would have difficulty, if only because power produces the subject and shapes its interest and identity' (Gordon 2002: 135). Here again we can distinguish Arendt's understanding of power as acknowledging the agency of subjects as able to change structures, to bring about something that is new and unknown – for better or for worse – while in the Foucauldian account agency is limited to always operate within the competing structures of shifting powers at the micro and macro level.

Arendt beyond Foucault

While calling attention to the symmetries in Arendt's and Foucault's work on life-politics in modernity is helpful for situating Arendt's work as relevant in the general discourse on biopolitics, it is the asymmetries and differences that are perhaps most telling in where the discourses of biopolitics might benefit from looking at Arendt's perspective in this context. Foucault has provided a powerful lens for scholars to consider recent international contexts of conflict, specifically the pugnacious practices instituted by the United States' administration since 2001 in the fight against terror. By investigating the liberal paradox in terms of biopolitics, he has certainly been instrumental in providing a novel perspective for looking at institutions and practices in terms of the powers they wield and the structures and subjects they wield it on. But the Foucauldian account also, naturally, has its limitations. Some scholars have raised the critique that his theories rely on an 'outdated vision of contemporary technology' (Braidotti 2013: 117), others have noted that neither is the issue of sovereign power fully resolved in Foucault (Lazzarato 2002; Jabri 2006) nor is his assessment of security in a biopolitical modernity fully developed (Dillon and Neal 2011: 11–12; Bigo 2011: 93–5). To look solely to Foucault for a comprehensive and fully coherent analysis of the contemporary biopolitical condition would be limiting our ability to fully understand the multiple ways in which life and politics are interwoven today. The codification of biopolitics is under way, but by no means complete yet, and this means that we should look beyond Foucault for insights into this condition (Campbell and Sitze 2013b: 1–2).

Arendt's inquiry into the subject of the biopolitical human condition is comprehensive in approach but no less detailed in the minutiae than Foucault's account, albeit perhaps slightly more polemical in tone. This often leaves Arendt open to criticism as she does occasionally make sweeping statements and constructs some tenuous relationships that require a benevolent attitude and a good dose of the benefit of the doubt granted by her readers. In a direct comparison to the Foucauldian approach of the analysis of biopolitics, Arendt's work does have some limitations. Roberto Esposito points towards a blind spot in Arendt's work, which disallows her to connect the material aspects of biological life to an authentic political sphere (Esposito 2008: 150). I agree with Esposito's assessment that the categorical exclusion of the materiality of human life from politics produces tensions in an Arendtian account of biopolitics; I propose, however, that these tensions can be eased by following Arendt's reasoning of the strict delineations she makes for what conditions are required for humans to be free and able to engage in politics. The next chapter engages in more detail with this question. However, when it comes to addressing overall materiality of biological life as it relates to the body and its functions specifically, Esposito has a strong point: Arendt did not engage satisfactorily with the body as a political site at all, and this is perhaps Arendt's gravest limitation for a comprehensive discussion of life politics. Overall in her writings, the body as a relevant political entity receives very little attention by Arendt. She confines the physical husk and interior processes of the human to the private sphere, as the 'master signifier of necessity' (Honig 2006: 360). As such, the body is not an object open for political discussion. This is a jarring assertion for any contemporary liberal viewpoint. Too pressing are political issues concerning gendered identity politics, for example, for this perspective to be palatable. However, if we stay faithful to the Arendtian account, it becomes clear why she excludes the body from political consideration. For her, the body has limitations as a form of political expression. Moreover, when the attributes of the body become politically relevant, there is a great danger for racial discrimination in Arendt's account. For Arendt, the body is not something that we can form or mould, it is something of a given – in political terms. Biological factors of the body cannot be altered or amended by debate, thus they must remain on the outside of politics proper for Arendt. We can then perhaps see Arendt not as overlooking the body as a political entity but rather as highlighting the fact that in any *political* inquiry the body has intrinsically limited scope.[5] Given this perspective, the problem of the body as an apolitical entity cannot be resolved in an Arendtian biopolitics.

So how can we then understand the non-material contribution that an Arendtian account makes to the discourses of biopolitics? And what does Arendt have to say that we have not already heard from Foucault? The crucial insights an Arendtian framework of biopolitics allows are threefold. Firstly, Arendt's account allows us to examine the consequences of biopolitics in modernity by looking through her lens of plurality. When we consider the pivotal role plurality plays in human relations and political

action, we are able to understand the ramifications of a modern political context that is centred on the biological mandate of bodily needs,
where commonality in the demand to have needs met is understood as
a shared political interest and the sameness of the body is taken as the
basis for equality (Arendt 2004: 200). Commonality of need, and the sameness of biological processes are emphasised in an Arendtian biopolitical
modernity, and difference, the basis for plurality, is diminished (Honig
2006: 361). This is amplified through technological means and perspectives that shape and inform ideas of the human as belonging to the category of a (biologically) functioning being. Through an account of Arendt's
life politics we can understand the dangers of this perspective. Arendt
warns of the implications of such an overstretched emphasis on commonality in which survival and progress of the species becomes the master
narrative:

> The monolithic character of every type of society, its conformism, which
> allows for only one interest and one opinion, is ultimately rooted in the
> one-ness of mankind. It is because this one-ness of mankind is not fantasy
> and not even merely a scientific hypothesis ... that mass society, where man
> as a social *animal* rules supreme and where apparently the survival of the
> species could be guaranteed on a world-wide scale, can at the same time
> threaten humanity with extinction. (Arendt 1998: 46)

The next chapter addresses these problematic modern convergences of
sameness and difference in a biopolitical modern context in greater
detail.

The second distinctive insight we gain with Arendt is an understanding
of the naturalised conception of politics and its consequences. Not only
are we humans determined, shaped and influenced by the predominance
of biological processes as paradigmatic for that which concerns politics as
management but we are also deeply embedded in societal-political structures that mirror closely the cyclicality of the life processes. While the
original meaning of biopolitics (before the Foucauldian turn) placed more
emphasis on the conception of the state as an organism, this relationship
has largely gone astray. With Arendt, we are reminded how thoroughly
the notion of life, of organic thought, is presented in the processual and
progressive structures of political society in modernity. When we recognise that modern understandings of politics rest on this strand of biopolitical perception, contemporary narratives that weave in organic metaphors
with political aims resonate most clearly. John Keane highlights this as
follows: 'Within democracies, medical metaphors sometimes also surface,
as when politicians speak of surgical strikes, sanitary cordons, mopping-
up operations and fighting the "cancer" or "plague" of terrorism' (Keane

2004: 2). Although Keane mitigates his statement by adding that mature democracies have moved beyond this euphemistic metaphor in their political affairs, it is a narrative that we recognise today as prevalent in discussions, debates and justifications of violence in the fight against terror in the contemporary context. This medical narrative indeed represents one of the manifestations of the *zoe*fication of politics, as I will delineate in greater detail in the chapters that follow.

When humans' understanding of self, as members of a species, and their political environment are predominantly based on the functionalities of life processes, survival becomes the master mandate. In the dichotomous relationship of life and death, death, as the antithesis to the life that must be secured, becomes the gravest danger. Not unlike in the Foucauldian account, Arendt lucidly recognises thus that the result must be a political focus on securitisation (Arendt 2003b: 443). And, as explained in the previous chapter, it is in this survival and progression mandate of the population as a whole in its totalising aspect that terror can unfold.

Where for Foucault the very basis of this inclusion/exclusion dichotomy of who can and must be sacrificed is racism, in the broadest sense (Foucault 2004: 255), for Arendt, this is a much more inclusive danger that potentially comprised all types and kinds of persons who, under some ideological strand or another, pose a danger to the progressive development of humanity, where foes of humankind are singled out as the 'objective enemy'. In Arendt's words: ' "guilty" is he who stands in the way of the natural or historical process which has passed judgment over "inferior races", over individuals "unfit to live", over "dying classes and decadent peoples" ' (2004: 599; Villa 1999: 185). Where the survival and progression mandate reigns supreme, everyone can become a potential threat in the perpetual struggle for progress.

The third pertinent insight we can gain from an Arendtian biopolitical framework is the modern perspective that perceives the human as essentially self-constituting, as a process of history and nature. In other words, humans in their scientific endeavour have come to see life (in manipulating life processes) as both human-influenced and human-controlled. Ray Kurzweil's vision of a 'singularity' in which we all become machine beings that can be improved, prolonged and replicated is the logical, contemporary escalation of this notion (Kurzweil 2006). Dana Villa traces this perspective back to Arendt's analysis of the 'hubris of homo faber' in the project of 'fabricating mankind' based on 'humanity's capacity to mimic and exploit natural processes' (Villa 1999:185). Villa identifies Arendt's realisation of a biopolitical turn in this as he notes that the project of fabricating humankind 'consists of the violent reshaping of available human material so that, in the end, neither classes, races or individuals exist, but only specimens of the (perfected) species' (1999: 185). In a dissociative turn, humans are at once maker of life and reduced to their (inferior) biology in 'making better'. Arendt's insights into the implications of the scientific turn in which both nature and humans had become subject and object of analysis for the fabrication and replication of biological life allows us to critically approach issues of post-humanism and the rise of artificial life, robotic life, as an improvement on humans and their capacities. Simultaneously, and here Arendt again directly raises a critique against Marx, the modern human understands history as something that is 'made' by humans (Arendt 1967). Both – the production of biological life and the production of history – require a reduction of 'the incalculable', the aleatory elements inherent in life, to something that is ascertainable (Villa 1999: 185); both require levels of control that cannot but seek to eliminate uncertainty, thereby creating the very category of uncertainty and risk. Herein lies perhaps the greatest peril of biopolitics: the demand to fully control life in an ever-increasing effort to eliminate its inherent uncertainties, specifically when paired with a narrative that claims the inevitability of certain natural or historical processes. The clamours of Silicon Valley about the inevitable technological progress for humanity faintly echo here. For Arendt, such determinism is perilous. She relates the extremes of such efforts with terror and her writing in *The Origins of Totalitarianism* is incisive as she states that 'terror seeks to "stabilize" men in order to liberate the forces of nature or history' (Arendt 2004: 599) I elaborate this in greater detail in the following chapter.

Conclusion

In this chapter I have sought to situate Arendt more firmly in the discourses of biopolitics by highlighting not only where she is in congruence with a Foucauldian perspective on biopolitics, but furthermore pointing out where she differs from and exceeds the Foucauldian account. The discrepancies between Foucault and Arendt are not surprising given the fact that they each had a different analytical focus. Where the production of knowledge and power is at the heart of Foucault's analysis, Arendt is chiefly concerned with the conditions for politics. Where Foucault primarily focuses on individuals and populations becoming both subject and object of politics, Arendt recognises that the political processes themselves are informed and underpinned by biological imageries. Lastly,

Arendt allows us to engage with a specifically shaped human subjectivity in more than one ways. The Arendtian human, though biopolitically shaped through a scientific and technologised understanding of life, retains a certain level of agency. An Arendtian perspective of life politics allows us to look at aspects beyond Foucauldian insights into the objectification of subjects through biopolitical technologies. It offers a trajectory that helps us to identify how humans, their politics and their ethics are biopolitically informed in a contemporary context and how this shapes in turn the sociopolitical subjectivity, but also how it need not be so. An Arendtian framework of biopolitics allows us to trace the implications of life-centric political practices for analysable, quantifiable, functional and controllable humans and raises questions as to what conception of politics a biopolitical politics facilitates. In her own analysis, Arendt makes the anti-political consequences of this totalising life politics perspective thoroughly clear and she provides us with an apt insight into the relationship between life politics and a decreased capacity for political action. It is the task of the next chapter to engage with this problematic in greater detail.

Notes

1 See for example Agamben 1998; Allen 2002; Barder and Debrix 2009; Birmingham 1994; Blencowe 2010; Braun 2007; Dolan 2005; Duarte 2006; Esposito 2008; Enns 2007; Fassin 2009; Gordon 2002; Hanssen 2000; Oksala 2010; Orlie 1997; Taylor 2007; Vatter 2006 and others.
2 Surprisingly, Foucault, in elaborating his theories on biopolitical techniques and *dispositifs* in modernity, remained virtually silent on the political theories of Arendt, who wrote her most salient analyses a good two decades prior to Foucault and who, in many respects, had already engaged with some of the biopolitical realisations in modernity, as the previous chapter has aimed to highlight. It can be assumed that Arendt's work was available to Foucault at the time (Bertani and Fontana 2004: 287); however, the reason for his omission of a deeper engagement with her scholarship can only be guessed at. Bertani and Fontana suggest that this is largely due to Foucault not keeping records of which books he had read and being furthermore not particularly interested in the polemics of an exchange with individual authors (Bertani and Fontana 2004: 287). Agamben, on the other hand, suggests that Foucault's study of biopolitics without any reference to Arendt is testament to the difficulties inherent in thinking about these matters (Agamben 1998: 4). Whatever the reasons, we can perhaps presume that Foucault was not entirely ignorant of Arendt's work at his time of writing and some of the parallels between Foucault and Arendt might well stem from a familiarity which the French philosopher might have had with Arendt's political theories on the structures and challenges of life politics in modernity, paired with a shared background of common influences.

3 It is worth highlighting that, unlike Arendt, Foucault refrains from addressing the historical development of the prevalence of the mathematisation of life that essentially enables this analysability of the human body.

4 See for example Allen 2002; Braun 2007; Blencowe 2010; Dolan 2005; Gordon 2002; Hayden 2009; Lupton 2012; Mbeme 2003; Swift 2008.

5 See for example Honig 2006; Tamborino 1999: 172.

3

Anti-political (post)modernity

A necessity ... is precisely what politics is not. In fact, it begins where the realm of material necessity and physical brute force end.

Hannah Arendt, *The Promise of Politics* (2005: 119)

The previous two chapters have considered Arendt's work as contributing to a biopolitical lens with which we can better understand our contemporary modernity. I have placed her in conversation with Foucault, so as to identify more clearly the dimensions we gain by a close reading of *The Human Condition* and other texts as relevant to the biopolitical turn. Specifically, I have traced the relevance of her account of biological processes as both the focus and model for modern politics and the scientific-technological mindset that underwrites these political forms. This and the next chapter discuss what an Arendtian life-politics turn means for both the political and violence in modernity. In this chapter, I argue, with Arendt, that, where life is the pivotal concept for political structures, the possibility for political contestation is problematically narrowed. In Chapter 4, I will show how the narrowing horizons for political contestation open up widening pathways for instrumental violence and its justifications as an expedient tool in the administration of life politics.

In her analysis of modernity and its impediments for political action, Arendt traces a coherent account of the shift of life into the centre of politics in two ways: as the focus *and* the basis for politics in the modern era. This *zoe*fication of politics and the concomitant politicisation (and technologisation) of *zoe* form the basis for the underlying tensions present in modern politics and political violence in more than one way. Biological life and political action are located for Arendt at opposite ends of the spectrum of the human condition. She sees biological processes as cyclical and 'given' by nature, by birth; they are thus mute as political constituents. Biology then is for Arendt a wholly inadequate basis from which to constitute politically active subjects. Chapter 1 briefly called attention to the possibility that there may be a perceived contradiction in terms when we consider the Arendtian categories of what belongs to and what should be excluded from politics in a biopolitical context. However, if we try and conceive of Arendt's biopolitics as an anti-political form of politics in modernity, it opens an avenue to further explore the implications and dangers of life at the centre of modern politics. And this, in turn, perhaps helps us understand why Arendt insists on her radical distinctions of private and public so as to preserve not only the possibility for political action but also the space and stage so vital for it. A first step in this investigation must thus begin with some basic but essential questions: why exactly can biopolitics not be included in what Arendt understands as politics proper but rather remains anti-political for Arendt? What are the dangerous implications she highlights in the emergence of a dominant anti-politics in the political realm? And how are we to understand the real-world practicalities in modernity of this strict categorisation?

In what follows, I focus on the implications of making the public and political realm one that has at its heart – and seeks to secure – the anti-political matter of biological life. I argue that, for a close reading of Arendt as a biopolitical thinker, it is important to distinguish between her understanding of what politics proper entails and how this stands in stark contrast to contemporary conceptions of politics-as-management. This then renders the biopolitical condition essentially an anti-political condition, as it impedes the fundamental aspects of what constitutes and facilitates political action proper. I trace this relationship by showing how fundamental aspects of politics are impeded, if not eradicated, through the biopolitical rationale and show, in the final part of the chapter, the relevance this has for the use of violence in a modern political context.

The (im)possibility of political biopolitics

Arendt has been widely criticised for her rigorous exclusion of matters concerning life necessities and the biological aspects of human life from the political realm.[1] Her resistance to treating physical bodies and their governance as political factors seems anachronistic in a time when large groups of people stand to suffer economically from a change in healthcare regulations (specifically in the US), when the rights of LGBT people constitute an important focus for political contestation and when women's reproductive affairs continue to animate fierce political disagreement. Indeed, in contemporary politics in which a key occupation of political governments is the distribution and management of resources that influence a wide range of life necessities in various ways, the Arendtian position of excluding matters of human biological life from political action appears untenable, if not outright delusional. However, to dismiss Arendt entirely

as not giving enough credence to the political relevance of the matters of
the body means that we overlook the nuance with which she treats both
life and, importantly, *politics proper*. I argue that if we consider her perspectives on politics and life necessities as radical positions on the limitations of both biological life and authentic political engagement among
humans, it becomes clear why a conflation of the two renders politics
proper in modernity precarious and dangerous.

For the purpose of clarifying Arendt's position, Andrew Schaap's (2010)
critique of Arendt's limited understanding of the historical-political implications of 'need' provides a useful point of departure. In his critique of
Arendt's anti-political perspective of needs, Schaap identifies her assessment of politics to include five key aspects (initiatory, constitutive, inclusive, performative and disclosive). He summarises the Arendtian position
on the exclusion of life necessities by juxtaposing features that exclusively
relate to the private, non-political realm with aspects that form the essence
of politics for Arendt. These contrasts include for Schaap the anti-political
priority of the collective life process in modernity, in contrast to the political initiation of an event; the anti-political imagery of nature, in contrast
to establishing a common, shared world; the lack of difference in identical
biological needs, in contrast to a plural political condition; the image of a
social body in contrast to the contingencies of a polity; and, finally, instrumentality in contrast to praxis (Schaap 2010: 160). Schaap rightly notes
that the Arendtian category of *animal laborans* may be considered 'pre-
political insofar as politics ... is only possible to the extent that the needs
of the body have already been satisfied' (2010: 159). In a modern context
where life necessities and biological survival have moved into the centre
of what politics is predominantly concerned with, this relationship changes.
Here again, Schaap correctly observes: '[a] politics of need becomes anti-
political when public life is overwhelmed by economic concerns ... Need
cannot provide an organising principle for an authentic politics because it
is the opposite of freedom' (2010: 159). It is, however, precisely here that
Schaap levies his critique against Arendt as rendering matters of human
need unduly 'depoliticised', thereby reduced to their ontological rather
than historical dimension. In this she fails, according to Schaap's analysis,
to see that human needs are contingent in that they are a product of political organisation, emerging from a social context, and that they cannot be
equated with necessity.

For Schaap, the politics of needs, in their historicised dimension, can
indeed be conceived of as contingent, and it is in this recognition of contingency that Schaap sees the potential for a 'properly political politics of
need' (2010: 164). For him, it is the socialised aspect, which Arendt evades,
wherein he sees the possibility for a framework to establish an authentic

politics of need in the Arendtian sense. Where Schaap ultimately seeks to
rescue the politics of need from a blind-spot that he identifies in Arendt,
I use his point of departure to demonstrate why a politics of biological
necessity (in contrast to need) is, for Arendt, always anti-political, not
merely by definitional fiat but by the very consequences of shifting life into
a political centre. While Schaap's analysis provides a thought-provoking
perspective of how the politics of needs might be conceived, through
others and contrary to Arendt, in terms of a potential 'world-disclosing'
aspect, I suggest that a closer engagement with the central elements of
Arendt's notion of modern politics and issues of biological necessity as
anti-political is merited if we are to better understand her very differentiated critique of life-as-politics. In what follows, I unpack the implications
of key features of each, in dichotomous relation to the other: difference/
sameness, freedom/necessity and realisation/futility.

Social difference and political equality

Recall that, for Arendt, politics 'deals with the coexistence and association
of different men' (Arendt 2005: 93). At the very heart of Arendt's understanding of politics and political action thus stands a fundamental condition without which politics could not exist, that of plurality. Only in a
context of *different*, individually unique persons coming together to
organise themselves by finding commonalities can political consensus be
realised. This plurality, our inherently different stories, experiences and
contexts, forms the basis for political action, as we come together as political beings in the public sphere and act on our differences. This difference
grants that politics is never entirely static, but remains in flux, retains the
capacity for rupture and new beginnings. It is in this condition of a plurality of difference that equality is granted as the result. When difference is
understood not as a facet of the infinite plurality of human beings but as
a threat to the unity of a normalised standard for humans as a species,
this fundamental condition of plurality is jeopardised. Arendt's critique of
the social realm, where 'necessity, not freedom rules the life of society'
(2005: 149), provides an insight into her crucial differentiations between
difference, multitude, equality and sameness. In Arendt's conception of
the social and the political realm, equality *and* difference play an important role. It is only in *equality* that people can come together and effect
political action in temporary cohesion, but it is only through *difference*
that plurality exists as 'not only the *conditio sine qua non*, but the *conditio
per quam* of all political life' (Arendt 1998: 7). A tension in the public
sphere, comprising both social and political domains, arises if either the
capacity for action is compromised or the condition of plurality is

jeopardised. This is most emphatically evident in Arendt's assessment of the implications of the French Revolution, when unmet biological necessities became a politicised *want*. It is important to stress here that Arendt does differentiate quite deliberately between poverty and want in this context when she critiques the bursting on the social and political scenes of 'dress, food and the reproduction of their species' (Arendt 2006a: 50). She contrasts this with the American Revolution, where poverty was also a problem, but where 'misery and want' were absent from the scene (2006a: 58). It is this condition of bringing a want (based on an urgent need) into a public realm that has the capacity to homogenise an otherwise plural multitude. And here the blurring of private needs with political solidarity begins for Arendt. In her words: 'the political trouble which misery of the people holds in store is that many-ness can in fact assume the guise of oneness, that suffering indeed breeds moods and emotions and attitudes that resemble solidarity to the point of confusion' (2006a: 84). The political basis of difference is thus not guaranteed when political action is reduced to a 'given' commonality of biological life necessities rather than based on the plurality of differences on which we could act.

Moreover, what appears as a veneer for equality – the shared suffering based on a shared lack of provisions – is for Arendt deeply misleading as a foundation on which to build a *contestable* political structure. It is on the basis of the reduction of human activities to their common factor in society – biological life – that plurality is stymied, and difference becomes an obstacle to politics rather than a facilitator. In her analysis of the origins of totalitarianism Arendt first and most clearly develops her idea that the inherent tension in making natural facts a matter of politics presents evident cause for concern. Recall that for Arendt equality, 'in contrast to all that is involved in mere existence, is not given to us ... we become equal as members of a group on the strength of our decision to guarantee ourselves mutually equal rights' (2004: 382). Politics on the basis of human necessity assumes equalitarian inclusiveness on the biological basis of membership in the species 'human', rather than guaranteeing each other equal rights on an agonistic basis. And it is precisely when natural factors become the basis of political equality, and equality, as a concept, is shifted from a politically assured basis to a social concept, that it becomes dangerous for Arendt as 'society leaves but little space for special groups and individuals, for then differences become all the more conspicuous' (2004: 74).

It is when attributes that cannot be changed at will (such as skin colour, sex or the colour of one's eyes) come to be seen not just as socially relevant factors but as pertinent politically facets that difference becomes unproductive, as it cannot be acted upon. The figure of the alien is crucial here: '[t]he "alien" is a frightening symbol of the fact of difference as such, of individuality as such, and indicates those realms in which man cannot change and cannot act and in which, therefore, he has a distinct tendency to destroy' (2004: 383). She illustrates this loss of equality and the freedom to act with the example of a dark-skinned person, who, in a white-skinned community, is considered solely as a bearer of his skin colour; he has lost 'along with his right to equality that freedom of action which is specifically human' (2004: 383). This political delimitation, she says, can happen more widely 'to those who have lost all distinctive political qualities and have become human beings and nothing else' (2004: 383). In other words, when we politically appear and are perceived primarily as our biological features and needs, then our possibility for political action and contestation is compromised. In politics, it is a 'who', not merely a 'what', that appears in public (Arendt 1998: 179).

In her analysis of human rights, and their perplexities as they relate to humans and citizens, Arendt explains the problem of equality and difference further (2004: 341–84). Her crucial insight here is that equality is always a product of human organisation. But such organisation can be brought about only through the precondition of the coming together and agreeing in action of persons who are distinguished by their difference. This difference, as we have seen, only functions constitutively as such if it is not reduced to mere natural characteristics, for these are entirely unactionable and present a 'limitation of the human artifice' (2004: 383). In a modern context where the boundaries of the social and political realms are indistinct, where social matters are elevated to issues that are to be engaged with politically, matters of life necessities become political. And where matters of biological human life enter politics, for Arendt this is the realm of *doxa*, of opinion, they become subject to debate and discrimination. For Arendt, the inherent givenness of biological factors, be it necessity or biological characteristics, is not a matter of opinion and should not become a political question as such.[2] Furthermore, they should not serve as a pretext for debate for the granting of equality. Equality on the basis of natural indicators remains always precarious in a juridical sense. Arendt draws an important distinction here between difference, as a political basis, and discrimination, as a social reality.[3]

Equality requires, or perhaps demands, a dichotomy of 'normal' and 'abnormal'. In the social realm this is a given for the freedom to associate, and by doing so discriminate between different social groups to which one feels a sense of belonging. It is when this discrimination moves into the political realm that the most violent consequences become possible of a politics that has as its referent biological life features. She states this as follows: 'the more equal people have become in every respect, and the more equality permeates the whole texture of society, the more will differences be resented, the more conspicuous will those become who are *visibly and by nature* unlike others' (Arendt 2003a: 200, emphasis added). In a society where homogeneity, based on biological sameness and fuelled by the production and consumption cycles of natural and quasi-natural processes, becomes the cultivated norm, natural and visible differences have a greater potential to be rendered and managed in terms of normal and abnormal. It is the visible and audible appearance that engages as the political, whereas 'inner qualities and gifts of heart or mind are political only to the extent that their owner wishes to expose them in public' (2003a: 199). It is precisely here that Arendt judges the admirable American principle of equality to have its greatest weakness, as it is 'not omnipotent; it cannot equalize natural, physical characteristics' (2003a: 200). Equality demands that each individual is considered as equal – every self another self – thereby creating a tension between groups that, for whichever reasons based on differentiation, are unwilling to grant each other such equal rights. The most visible of these differentiation criteria, or markers, are of biological nature, such as colour of skin and difference in features.

The social-political equality that different-but-equal persons grant one another has thus limits, and these limits are what effectively then constitutes a biopolitical ground. The violence that resides in this arrival of biopolitics is clear when we consider the unactionable and delimiting nature of a reduction to the biological. Equality as a fundamentally political principle, as based on the capacity of action, is thus limited, and, in its limits, limiting. She explains:

> Equality of condition, though it is certainly a basic requirement for justice, is nevertheless among the greatest and most uncertain ventures of modern mankind. The more equal conditions are, the less explanation there is for the differences that actually exist between people; and thus all the more unequal do individuals and groups become. (2004: 74)

Arendt speaks here of an 'equality of condition' that provides a basic requirement for justice but she states: 'When equality becomes a mundane fact in itself, without any gauge by which it may be measured or explained then there is one chance in a hundred that it will be recognized as a working principle of a political organisation in which otherwise unequal people have equal rights' (2004: 74). The closing of a space for difference, the mandate for a universally equal public realm renders biological, physical and also cultural differences a distinct anomaly within the public realm. And it is here that Arendt's diagnosis of an increasingly conformist society that responds positively to the normalising practices and discourses of biopolitics gains relevance.

Plurality in the public realm is, as is well known, the very cornerstone of the *vita activa* for Arendt. Only if we are to understand the social realm as a pre-political condition does her argumentation remain in line with her priority for plurality. A social realm dominated exclusively by the sameness engendered by matters of life necessity becomes thus anti-political. This then becomes the most treacherous realm in modernity as, in its extreme potential for conformism, difference is always in danger of becoming diminished, leaving those natural attributes that cannot be made to 'conform' to an obvious parameter for inclusion or exclusion practices in societies where life, and the unfettered functioning of the production and consumption process is the highest aim. It is when life, its necessities, its immutable characteristics becomes the primary common denominator that the spaces for trans-natural difference is endangered.

Moreover, in a political sphere in which 'political bodies are based on the family, conceived in the image of the family' (Arendt 2005: 94), and ideas of kinship form the basis of political organisation, Arendt detects nothing but the downfall of politics: 'in this form of organisation, any original differentiation is effectively eradicated in the same way that the essential equality of all men, insofar as we are dealing with man, is destroyed' (2005: 94). Not only is the realm of the family that in which the security of the life process is the central priority (Arendt 2003b: 448) but the image of the family, as the basis for politics, results in a quasi-divine aspiration to create, and intervene in, the concept 'human', 'by acting as if we could naturally escape from the principle of human differentiation' (Arendt 2005: 94). In seeking to establish a concept of kinship that attests to the 'sameness' of all who are to be included in the concept 'human', biology is the most obvious common, and lowest denominator and the human is fundamentally reduced to a member of the *species* human, whereby the survival and perpetuation of the life processes of members of the gigantic family of humankind are sought to be secured. For Arendt, this ultimately represents a perversion of the political as it prioritises kinship over plurality (Arendt 2005: 94). And in this quandary she identifies the dangerous and anti-political basis of modern (Western) politics:

> The west's solution for escaping from the impossibility of politics within the Western creation myth is to transform politics into history, or to substitute history for politics. In the idea of world history, the multiplicity of men is melted into one human individual, which is then also called humanity. This is the source of the monstrous and inhuman aspect of history, which accomplishes its full and brutal end in politics. (2005: 95)

The escape from the freedom of politics into the necessity of history was, for Arendt, a reprehensible incongruity, and its implications for politics, for the world, dangerous. Political procedures that are based on the family and the protection of biological processes delimit this essential plurality and thereby render such types of political government anti-political. Political action is thus severely delimited in its constitutive capacity to establish, through non-biopolitical politics, a constitutional framework, in the Arendtian sense, that could perhaps, at a social level, comprise solutions to difficult questions of resource distribution and related issues.

Collective *philopsychia*

Following Arendt's view of what constitutes the political, it is easy to see that modern and late modern conceptions of politics only faintly meet the conditions required for politics proper. In the (perhaps overly ambitious) contemporary mandate for universal equality, based on a notional concept of humanity, it is assumed that the pre-existing condition of being a member of the human species implicitly determines a universal (human) right (Balibar 2007). On both a national and international level, it is evident that this has to date not proved to be a viable deterrent for violence; on the contrary, it seems to engender more violence in the name of restoring an abstract ideal of humanity. It is perhaps helpful to consider the precarious politics of abstract universal equality in a contemporary context to understand the implications that Arendt sought to highlight. Here, in a more stringent appropriation of the agonistic and antagonistic realities in politics, Chantal Mouffe makes a case for the normative consideration of the political as encompassing the potential for antagonism inherent in every political exchange. While Mouffe and Arendt speak from different perspectives of what constitutes the political as such, both accounts depart from the supposition that the political requires plural perspectives as a basis.

In her discussion of the political, Mouffe criticises the prevalence of rationalist, universalist and individualist political democratic thinking that essentially stems from an '[e]nlightenment universalism of an undifferentiated human nature' (2005: 13). Therein resides a major flaw for Mouffe: such a position remains blind to the specificity of the political in its dimension of conflict or decision – in short – antagonism. Hence, a universalising politics is 'fraught with danger since it leaves us unprepared in the face of unrecognised manifestations of antagonism' (2005: 2). For Mouffe, antagonism, manifested through the differences inherent in a plural body politic is as much a social and political reality as discrimination and difference is for Arendt. In her advocacy of radical democracy Mouffe suggests: 'when we accept that every identity is relational and that the condition of existence of every identity is the affirmation of difference, the determination of an "other" that is going to play the role of a "constitutive outsider", it is possible to understand how antagonisms arise' (2005: 2). The moment of recognition of difference is important here as it is also the very moment that contains the potential to avert the antagonistic manifestation of actual political conflict and to overcome an existentially relevant inclusion/exclusion distinction by channelling it into agonism, based on mutual respect (equality for Arendt). Echoing Arendt's insistence of plurality as the fundamental condition for political interaction, Mouffe emphasises the dangerous implications when opposing perspectives (which always already exists) arise in a collective that identifies difference with a dichotomous 'we' versus 'them' lens.

For Arendt, humans associate *socially* on the basis of their differences; here a discriminatory 'we/them' mindset is not a problem for her because it is a non-political association. It is when the we/them construct becomes a political friend/enemy decision, in the Schmittian sense, that the non-political basis of social discrimination can turn into existential political differentiation. Mouffe explains: 'This can happen when the other, who was until then considered only under the mode of difference, begins to be perceived as negating our identity, as putting in question our very existence' (2005: 3). In other words, it is through the us/them dichotomy, which is established through collective identity (or in Arendt's terms social 'discrimination') – not individual identity (in Arendt's terms 'difference') – that the potentially problematic Schmittian political friend/enemy distinction emerges. In a biopolitical friend/enemy context the risk of political exclusion on the basis of survival is latent, the possibility for exclusionary violence elevated, and the possibility for political contestation is diminished.[4]

When the 'other' is designated an enemy who jeopardises the primary political project, namely life and its progression, their elimination must follow. When they are considered an adversary, 'whose existence is legitimated and must be tolerated', mutual equality as the basis for agonistic and pluralistic politics can ensue (Mouffe 2005: 4). Mouffe, via Carole Pateman, effectively illustrates the discriminatory potential of political equality that is granted on the basis of an abstract concept of membership in the human species. Pateman suggests:

> The idea of universal citizenship is specifically modern, and necessarily depends on the emergence of the view that all individuals are born free and equal, or are naturally free and equal to each other ... We are all taught that the 'individual' is a universal category that applies to anyone and everyone, but this is not the case. 'The individual' is man. (Pateman 1986, cited in Mouffe, 2005: 13)

A universal equality that rests solely on the basis of membership in the human species offers thus no guarantee for actual political equality in matters of politics when social norms based on conformity appropriate the realm of political action. As the modern social realm, characterised by 'cooperative, identity-destroying labour' assumes the role of the political sphere, which in Arendtian terms is marked by 'competitive, identity-disclosing action' (Bull 2005: 678), then matters of biology and life processes and necessity enter a field of competition, in which issues of life and survival are at stake and biopolitical forms of political violence appear as a means to remedy the cohesion of the abstract species: humanity.

Where humanity is the political master narrative of a sociopolitical body that has as its chief purpose the security, promotion and progression of the life processes of the species, survival becomes the key purpose for such a political society. And it is this turn, in politics becoming the administration of life in modernity, that politics substantially loses its meaning for Arendt: 'if it is true that politics is nothing more than a necessary evil for sustaining the life of humanity, then politics has indeed begun to banish itself from the world and to transform its meaning into meaninglessness' (2005: 110). Here, once again, the antithetical nature of the constellation of life necessities and politics becomes evident. It is without question for Arendt that politics proper means neither protection nor security, but its very meaning is freedom. Recall Arendt's assertion that 'in politics not life is at stake but the world' (2003b: 448). Liberal modernity no longer fulfils this condition for Arendt, as all action is governed by necessity in an 'increasing sphere of social and economic life whose administration has overshadowed the political realm ever since the beginning of the modern age' (Ibid.). The perpetual maintenance of life processes and the predominance of survival as the prime focus, not only of private but now also of public concern, renders 'man' in the singular and 'men', collectively fundamentally unfree in the public realm. In liberal modern government, this is reflected no longer in the Hobbesian security of the individual against death but rather a 'security which should permit an undisturbed development of the life processes of society as a whole' (Ibid.). In Arendt's terms, liberal modernity thus reflects more closely the communal actions and codes for conduct of a non-political *tribal* association than a body politics proper in its dominant concern for life processes and the lack of space for both politics and freedom, as she claims that it is those communities that do not form a body politic in which 'factors ruling their actions and conduct are not freedom but the necessities of life and concern for its preservation' (2003b: 442). In a modern society in which political preoccupations predominantly centre on matters of human survival and the progress of humanity, or at a minimum on the security of biological human life, the maintenance of the world, which provides for Arendt the spaces for people to come together in their agonistic plurality and act contingently in concert, loses meaning entirely. In other words, when in politics life is at stake, security becomes a key political paradigm and all political resources that must necessarily focus on this very condition become anti-political as such.

To understand the impasse at which politics in modernity has arrived for it is useful to review briefly Arendt's conceptions of both necessity and freedom and the consequences they bear when politics and necessity become conflated. Freedom and necessity are not only located at opposite ends of the spectrum but are for Arendt mutually exclusive: '[a] necessity … is precisely what politics is not' (Arendt 2005: 119). Here, another Arendtian premise comes to bear. For individuals to be politically active they cannot be preoccupied chiefly by matters of private, biological life – 'like hunger, or love' (2005: 119). Rather, to engage in politics proper, a person must emerge from the realm of life necessity, which by definition is always unfree. Freedom and politics thus are reciprocally related. Arendt does not understand freedom as 'being free from' something, but rather as a condition that we can experience only as intersubjectivity, in the interaction with other (free) individuals. The term freedom, in the modern context, has become somewhat ambiguous and the meaning of freedom today is somewhat obscured. In a contemporary Western context, freedom more often than not denotes 'freedom to buy', or economic freedom, and, particularly in the past decade, has been linked closely with the idea of humanity.[5] Increasingly also, freedom is conceived of as being free from the inconveniences of life – a narrative that is particularly prevalent in the technology industries.[6] In Arendt's own expositions it is not always entirely clear what precisely the term freedom denotes as it is presented in various nuances throughout her work. What is clear, however, is that the modern discussions of freedom,

> where freedom is never understood as an objective state of human existence but either presents an unsolvable problem of subjectivity, of an entirely undetermined or determined will, or develops out of necessity, all point to the fact that the objective, tangible difference between being free and being forced by necessity is no longer perceived. (Arendt 1998: 71)

Arendt critiques the modern understanding of freedom understood as 'a river flowing freely' (2005: 120) in which every attempt to block its flow is an arbitrary impediment. Where a kinetic idea of freedom prevails, the contrasting opposite to freedom is no longer necessity, but rather 'arbitrary action' or contingent political action. A flow is always determined by directionality, which in life politics translates into a normative and deterministic conception of how the human or humanity *should* progress within this flow of history. A directional flow narrows the space and scope for the unpredictable, the chaotic, the contingent as such. The dangers of a constellation where the free 'flow of history' (2005: 121) is equated with politics and takes place within a totalising sphere of life politics were clear to Arendt:

> The distinction between such pervasive ideological thinking and totalitar-
> ian regimes lies in the fact that the latter have discovered the political means
> to integrate human beings into the flow of history in such a way that they
> are totally caught up in its 'freedom', in its 'free flow', that they can no longer
> obstruct it but instead become impulses for its acceleration. (2005: 121)

Where humanity is understood as deterministic, as process-determined, progressive in nature, that which is not considered conducive to the historical flow of the evolution or progression mandate must either be absorbed or be eliminated. Survival, then, is the trophy gained in the evolution contest, but it is an unstable and fluid prize as the inherent contingency of human action renders the teleological attainment of a universal humanity impossible.

Contrary to these modern perceptions, freedom, in Arendt's account, means originally the freedom to leave the private realm of the household. In other words, freedom, in part, meant liberation from the realm of the family in which survival and the taking care of life's necessities was ensured. But this was by no means a passive act. For Arendt, leaving this realm, in which one always stood necessarily under the sway of necessity or coercion, required what has become one of the earliest political virtues: courage. She states: 'thus only that man was free who was prepared to risk his own life, and it was the man with the unfree and servile soul who clung too dearly to life' (2005: 122). Arendt uses the Greek term *'philopsychia'* in this context to denote the unfree individual who had a 'love of life for life's own sake' (Arendt 2002: 287; 2005: 122). To put it perhaps more bluntly, *philopsychia* characterises those persons who lack the courage to have any other ambition than to engage in labour as their sole activity. Modern society is, in Arendt's terms, fully under the sway of life necessities, whereby all political activities have been subsumed as quasi-politics in administrative tasks of life management. In such a society, *philopsychia* indeed appears as the collective psychological organising principle. Politics proper relies on action, and action, for Arendt, is closely related to risk. Where life is at stake in politics and survival becomes its paradigm, fear is the emotional driver of political actions. The consequences of a politics of fear have transpired and continue to transpire in the context of the US-initiated 'with us or against us' rhetoric in the past decade and illustrate effectively the dangerous margins of policies based on an emotion.[7]

There is a further aspect that must be considered in the discussion of anti-political biopolitics. It relates to the modalities with which politics is realised and which the labouring process precludes: the possibility of speech and language as a means to express politics. Arendt sees the labouring process, which is that process that relates to life necessities, as an inherently *mute* process. She explicitly relates matters of necessity to a most private, carnal experience that is characterised by its very incommunicability – the experience of pain. The outward manifestations of labour and of taking care of life necessities essential to survival are closely associated with pain and effort (Arendt 1998: 135). Likewise, when life necessities are not met and the body suffers deficiencies, pain in its various physical substantiation is the result. And pain, as Arendt notes, is one of the most world-denying and all-consuming realities possible. In her words:

> Indeed the most intense feeling we know of, intense to the point of blotting out all other experiences, namely the experience of great bodily pain, is at the same time the least communicable of all. Not only is it perhaps the only experience which we are unable to transform into a shape fit for public appearance, it actually deprives us of our feeling for reality to such an extent that we can forget it more quickly and easily than anything else. (1998: 50)

The crucial point here is that the experience of pain, while an expression thereof can be approximated in similes, can never be shared or fully communicated in a public realm. This is a serious impediment to political action because it is through speech, through communication, that politics materialises for Arendt. A society that stands under the perpetual duress of labouring and 'making a living' is a society under the sway of pain and effort and thus is a mute society. It is for this reason alone thoroughly unsuitable as a basis for politics, in Arendtian terms. If we accept her assessment that necessity and, derivatively, labour manifest as pain and effort, and that the lack of necessities met results in the very same, and we acknowledge that pain is the least communicable of all experiences, then a politics of necessity, or biopolitics, in the Arendtian sense of politics proper, where people appear freely in public and organise themselves through speech and action, is categorically antithetical. And it is here that Arendt's view of the muteness of violence echoes most strongly, as she critiques Marx for conflating history and politics and making violence the 'facilitator' of history to come.

The matter is further complicated when we consider that Arendt by no means wished to do away with necessity; on the contrary. Recall that labour is a fundamental activity in the human condition. Therefore freedom from necessity is inherently impossible, as necessity is as intrinsic a matter of the human condition and so inextricably entwined with life, that to 'free man from necessity', as was the Marxist aim, is fundamentally impossible. For Arendt, 'life itself is threatened where necessity is altogether eliminated' (1998: 71). Freedom *from* necessity, although a condition for being able to freely participate in Arendtian politics proper, does not constitute the establishment of freedom. Necessity remains a constant for Arendt, as long as there is life, necessity is its shadowy companion. She reiterates the same point in her diagnosis of a modern society that has become a society of labourers and consumers, as she warns that, the easier these activities appear, the less aware we may be of the urges of necessity; not as they are not present, but because they do not manifest themselves in pain and effort. Neither do they manifest themselves in freedom or the capacity for political action as such. With technological applications and machineries easing the pain of labouring, hers is a timely caution. Quite the contrary – Arendt sees a grave danger in such a society that is caught up in its own radiant fertility and perpetual and cyclical processes that it is unable to recognise its own inability to contribute to making a shared and lasting world. In this, Arendt quotes Adam Smith when she critiques that such a society would 'no longer be able to recognize its own futility – the futility of a life which "does not fix or realize itself in any permanent subject which endures after [its] labour is past"' (1998: 135).

Arendt was under no illusions that the meaning of politics in modernity would remain that of freedom, and she understood clearly that the shift of life into the centre of politics could have taken place only once the meaning of politics had become altered. She also acknowledges that the duality of the emancipation of labour that could come about only with this shifted perspective of politics as she states: 'the emancipation of labor, both as the glorification of the labouring activity and as the political equality of the working class, would not have been possible if the original meaning of politics – in which a political realm centered around labor would have been a contradiction in terms – had not been lost' (Arendt 2002: 286). But it is also in this shift that one of Arendt's key criticisms resides with regards to the anti-political life politics in modernity: the confusion and conflation of categories she considers to be distinct elements of the human condition: labour, work and action and the resulting violent consequences, whereby life becomes an abstract end and the means to maintain life 'will become, over time, the end' (Young-Bruehl 2006: 56).

Means–ends, or the end of meaning

Arendt's critique of the conflation of the categories labour, work and action rests in large part on her critique of Marx, who put forth two fundamental and troublesome theories: 'First, Labor is the Creator of Man; second, Violence is the midwife of History' (Arendt 2002: 287). These propositions are problematic specifically against Marx's position that history is past political action, which in turn means 'violence makes action efficient' (Arendt 2002: 287). Such thinking introduces ideas of utility and functionality into the political arena, which, for Arendt, was wholly unhelpful. Specifically against a backdrop in which humanity is comprised as an abstract entity, where survival and progress serve as the primary social and political paradigm and where labouring processes are the key to survival, usefulness, purpose and efficiency become primary goals. With life as the highest good and labour elevated to the highest activity, metabolic considerations reign supreme: 'what was not needed, not necessitated by life's metabolism with nature was either superfluous or could be distinguished only in terms of a peculiarity of human as distinguished from other animal life' (Arendt 1998: 321). In an instrumental understanding of the labouring process, life is necessarily rendered instrumental. Action, and by extension politics,

> was soon considered and still almost exclusively understood in terms of making and fabricating, only that making, because of its worldliness and inherent indifference to life was now regarded as but another form of labouring, a more complicated but not a more mysterious function of the life process. (Arendt 1998: 322)

Arendt was concerned that the cyclical, repetitive and literally futile acts involved in all life processes were recast as 'creative' work, as a form of action, in modernity. Where labour subsumes work, the backdrop of a shared world turned from permanence to futility. Here futility is not an expression of pointlessness but, rather, expresses the cyclical structure of consumption and production which is contrary to building permanent structures, as is the concern of work in Arendt's account. The means–ends question so critical in Arendt's instrumental category of work is entirely meaningless where matters of life processes are concerned. With the management of life processes as the chief aim in a political community, and a perspective that deems history as a deterministic political process, it is not surprising that acts of politics, driven by utility, became not just literally but also metaphorically linked to the natural biology of life. The instrumentality paradigm is crucial to biopolitical modernity. It is precisely what liberates the use of violence in the conflation of life with politics. If life processes are absorbed into a Marxian process of 'making' history, then politics stands under the sway of an organic metaphor by which the end of history must be violently brought about. Arendt frequently criticised what she considered to be one of the most forceful metaphors underlying the justification of violence, namely that 'violence is linked to labour pangs' (Arendt 1967). Here, the *zoefication* of politics discussed in Chapter 1 is most evident and most perilous. The organic metaphor not only renders the use of violence as a political tool *zoefied* but also is for Arendt vastly misleading as it quite blatantly equated biological life with politics and suggests that violence can act as a viable means to fulfil history, a history that is linked with ideas of both the process and the progress of life.

The use of violence is for Arendt always instrumental and relies on tools and technologies for its executions; it has no language or speech, and as such no essence of its own. It exists in the political realm not independently but only ever through (verbal) justification. Arendt reminds us: 'what needs justification by something else cannot be the essence of anything' (1970: 51). She dismisses the efficacy of embodied and restorative violence. Where Frantz Fanon famously embraces the physical embodiment of violence in the anti-colonial struggle in his seminal work *Wretched of the Earth*, Arendt finds 'nothing, ..., more dangerous than the tradition of organic thought in political matters by which power and violence are interpreted in biological terms' (1970: 75). Violence conceived in terms that evoke the natural birthing process for the progressive development of a body politic that is concerned first and foremost with life processes is for Arendt a dangerous and whole anti-political proposition.

Her cautions against rendering violence as organic are persuasive in light of the totalitarian reality she herself experienced and which continues to cast its shadow far into modernity. The *zoefication* of politics – placing emphasis on the biological as the basis for the political – allows for the instrumental rationalisation of biopolitical violence, as Arendt explains:

> So long as we talk in non-political, biological terms, the glorification of violence can appeal to the undeniable fact that in the household of nature, destruction and creation are but two sides of the natural process, so that collective violent action, quite apart from its inherent attraction, may appear as a natural prerequisite for the collective life of mankind as the struggle for survival and violent death for continuing life in the animal kingdom. (1970: 75)

Aware of the temptations of such justifications in her own time, Arendt rejects the conflation of violence-as-politics with natural processes, which she, like Foucault, sees as a catalyst for racism. If politics and its violent instruments are defined as life processes, then the Schmittian friend/

enemy distinction becomes one that is based on what is identified as hostile to humanity, in the modern context dominantly manifest as an ideology of race distinction. Violence as a legitimised and justified instrument of politics here becomes the murderous tool for racial exclusion in the totalitarian context. And it is indeed in this type of conflation of the political with the biological and the sovereign monopoly on the rational use of violence that Foucault demonstrates the murderous biopolitical racism of the totalitarian regime. As he states: 'We have then in Nazi society something that is really quite extraordinary: this is a society which has generalized biopower in an absolute sense, but which also has generalized the sovereign right to kill' (Foucault 2004: 260). The glorification of violence as an essential and perhaps even necessary element of life is inherently antithetical to Arendt's understanding of what makes humans political. She states: 'Death, whether faced in actual dying or in the inner awareness of one's own mortality is perhaps the most antipolitical experience there is' (1970: 67). As such, it would seem contradictory for her to consider violence as anything other than an act that stands fully outside of the realm of the political. Chapter 4 elaborates this in more detail.

The collapse of the categories labour, work and action solidifies two changed perceptions. Firstly, that labour, as the prime activity of humans and necessary for the survival and progress of humanity, now constitutes our link to a shared community based on biological characteristics; and secondly, that politics is now, literally, seen as instrumental in facilitating the labouring process for humanity. The Marxian adage of 'violence as the midwife of history' fits seamlessly into this perspective and, if seen against Arendt's warnings, reveals the dangers inherent in understanding violence as necessary to secure humanity in a *zoe*fied understanding of politics. Following Arendt's analysis, it appears as though, for her, in modernity, there has been a shift in our capacities for collective political action from making a shared and lasting world to managing life itself as a literal and metaphorical biological process. No longer are we chiefly engaged in improving political life, bettering constitutional frameworks, enhancing capacities for politics proper; instead, there is a preoccupation with how we may have to labour less for our sustenance, improve biological functions of our individual bodies and eliminate actualised and perceived risk to human biological life.

The body has become a site of production and creation in the very literal scientific sense. *Homo faber*, the toolmaker in the Arendtian account, no longer fabricates the world, but produces tools and instruments to facilitate the life process. Soon, Arendt argues, machines to facilitate labour are what bind people to a shared world, eventually subsuming the labourer entirely in the technological operational labour processes (Arendt 1998:

149–51). Today we find ourselves at this junction. Today, such machines both mirror and produce biology and matters of life processes, whether that be in bio-molecular medicine or in warfare – the technological singularity that Arendt presciently was conscious of – is allegedly inevitably upon us, further thrusting humans away from the realm of politics proper in which human difference can be revealed, marginalising human political experience altogether. Humans in their biology fulfil a productive purpose. Michael Dillon and Julian Reid note that in modernity a pivotal question arises: *'What are people good for?'* – and it is precisely this question that renders biopolitics so precarious and anti-political. In their words:

> Biopolitics cannot abide the good for nothing. But politics, that process by which order is changed to accommodate new principles of order rather than to rank all principles under a common metric, must insist on admitting the good for nothing to the conversation; must admit the good for nothing to the conversation, because that is the vocation of the political. (2009: 154)

Arendt's perspective of what constitutes politics proper strongly echoes in this assessment. It is clear that for Arendt, in line with Kant, neither humans nor the life processes that sustain humans should be conceived in utilitarian and predominantly functional terms but rather as ends in themselves. Where utility is radicalised, 'all ends are bound to be of short duration and to be transformed means for some further ends' (Arendt 1998: 154). When violence is the means by which an end that relates to the security of survival and progress is to be achieved, the meaning of violence in society becomes subject to transformation.

Conclusion

The concept of an Arentian biopolitics is challenging. Her treatment of the body and its needs as pertaining to the hidden, non-political private realm is jarring for today's political conditions and in need of revision. Nonetheless, we should take seriously her sharp insights into the complexities and difficulties of biopolitics in terms of its debilitating impact on politics and its intrinsic potential for violence. The political human is for Arendt far more than a living being, it is a being that is woven into the social, political and physical fabric that constitutes the world. This human is bestowed with rights as a political being, not merely as a physical being. The latter subsuming the former posed a great danger for Arendt, not just to the possibility for politics proper but to the world at large.

When understood in the context of differing conceptions of politics, we can see how biopolitical politics is, in fact, anti-political in Arendtian terms, and cannot be included in what Arendt understands as politics proper. Politics conceived of as administrative or managerial processes of law, order and distribution differs considerably from an understanding of politics as a place for collective agreement and contestation, for the formation of political structures and a shared world. Arendt's politics proper is always open to contestation – even the structures that have been erected. However, where political action is no longer possible, where political structures obstruct it, politics proper, and with it contestation, is no longer assured. Such was her concern about the modern condition of life politics. Importantly, the two contrasting perspectives of politics have different theoretical implications for how we perceive and justify acts of violence in relation to politics in modernity, as the prevalent political perspective will 'invariably shape our answers to innumerable questions about what should be punished, when nominal violations are justified and when wrongdoing should be excused' (Fletcher 2000).

Life and politics as a conjoined concept is one that impacts on how we conceive of political action and political spaces in modernity. Difference, speech, freedom and the embrace of uncertainty and risk, vital categories for the possibility of politics proper in Arendt, disturb the biopolitical mandate of politics-as-management and put into question the zoefied processes of political administration that are reliant on control and predictable outcomes. Where heterogeneity is absorbed by homogeneity, language and speech are muted and freedom subsumed by necessity, politics proper is at risk. However, where the principles for politics proper are endangered, forms of violence, in lieu of politics proper, may more readily be employed in creating certain ends, and may, indeed, be glorified as the necessary instrument to make humanity survive and progress. In other words, a zoefied understanding of politics runs thus the risk of conceiving of violence in terms of its creative capacities for political ends, and can be justified in terms of necessity and control. The occlusion of the principles for politics becomes ever-more pressing when we consider the homogenisation mandate of new anthropo-technological developments (Campbell 2011: 115; Sloterdijk 2009: 24–5). The problems this produces are developed in Chapter 6. In the next chapter, my concern is the role of violence and its justifications in a biopolitically informed, anti-political condition. Here I will draw one final time on Arendt's astute insights into the life–politics–violence nexus.

Notes

1 Among other works, Gareth Williams's *Hannah Arendt: Critical Assessments of Leading Political Philosophers* (2006) provides a comprehensive collection of critiques of Arendt's theories. Specifically contributions from Richard

Bernstein and Jeremy Waldron deliver some interesting aspects on her perspective concerning the social and the political. (Williams 2006)

2 A brief digression perhaps illustrates Arendt's position on this matter. In a discussion at a 1972 conference organised by the Toronto Society for the Study of Social and Political Thought Arendt was pressed by her colleagues to engage with the criticism of why matters of social needs should be so rigorously excluded from political discussions when it seems so entirely impossible to separate social issues from political questions. Arendt illustrated her point as follows: 'Let's take the housing problem. The social problem is certainly adequate housing. But the question of whether this adequate housing means integration or not is certainly a political question. With every one of these questions there is a double face. And one of these faces should not be subject to debate. There shouldn't be any debate about the question that everyone should have decent housing' (Bernstein 2006: 247).

3 In her essay 'Reflections on Little Rock' Arendt states bluntly that discrimination is in fact a social right. Given that she has been somewhat vague with the status of rights in a social realm I am not entirely sure this formulation contributes to a constructive discussion and prefer to discuss the issue of discrimination in society in term of a social reality or fact rather than a right. It is important to see, however, that Arendt, in her categorical conception of the social, sees no problem with discrimination in the social realm just so long as it does not become political, which of course can no longer be ensured in modernity, according to Arendt.

4 As Andrew Schaap rightly points out, the conceptions of agonism in Mouffe and Arendt vary in that the former presents the agonistic exchange in light of the distinction between groups, while Arendt's perspective on antagonism is based on the infinite plurality of human beings. This does not, however, bear on the illustration of the argument presented in this chapter, but must be considered in the wider context of this book (Schaap 2007).

5 A telling example of the declaratory relationship of freedom with humanity is captured in the infamous 2002 US National Security Strategy document: 'Freedom is the non-negotiable demand of human dignity; the birthright of every person—in every civilization. Throughout history, freedom has been threatened by war and terror; it has been challenged by the clashing wills of powerful states and the evil designs of tyrants; and it has been tested by widespread poverty and disease. Today, humanity holds in its hands the opportunity to further freedom's triumph over all these foes' (White House 2002).

6 Particularly the world of app development and technological enhancements market their advancements and inventions by highlighting how their products ease the chores and inconveniences of life. Paradoxically, technological devices and systems such as Quantified Self technology, in which individuals can track, measure and control their biological life systems via sensors and computers as small as a wristwatch, 24–7, render life further calculable and controllable. In Arendtian terms, this is the antithesis of freedom. Rather, it draws individuals into an introspective process of prioritising life processes, potentially ad infinitum

7 The shift of practices, such as torture and renditions, that were hitherto considered to be absolutely prohibited in warfare, into a framework of policies that
essentially awards the president of the United States the capacity to allow such
practices (as initiated by George W. Bush and likely continued by Donald
Trump) or prohibit them (as in some of Barack Obama's policies) is just one
such example.

4

Procedural violence

Listen – there's no war that will end all wars.
> Haruki Murakami, *Kafka on the Shore* (2005: 416)

The biopolitical anti-politics of modernity is a dubious form of politics for Arendt, one that has ceased to have meaning, beyond meaninglessness, once it becomes nothing more than 'a necessary evil for sustaining the life of humanity' (Arendt 2005: 110), and this meaninglessness opens the pathway for violence as an expedient tool in the administration of humanity. Where the preceding chapter has shown how a biopolitical rationale shapes perspectives of politics-as-management, occluding politics proper, this chapter turns its task to showing how this, in turn, shapes the relationship between politics and violence. In the radical governance of biopolitical processes, violence does politics' bidding for the management of useful and useless life. Here, biopolitics becomes thanatopolitics (the political administration of death), as death is inevitably twinned with life, in that 'the decision of life becomes a decision on death' (Agamben 1998: 122). Where Giorgio Agamben uses the term thanatopolitics to denote the sovereign's extrajuridical reach into the decision of who is rendered killable for the prosperity of life, Achille Mbeme sees the 'subjugation of life to the power of death' as a form of necropolitics (the political administration of dead bodies) in which ever-growing populations are subjected to topographies of cruelty that render them living corpses in a biopolitical logic (Mbeme 2003: 39–40). In both accounts of the dark radical side of biopolitics – whether referring to totalitarianism or the contemporary production of 'death-worlds' (Mbeme 2003: 39–40) – violence is instrumental to biopolitics in quite a literal sense.

The *zoe*fication of politics provides the referent for a logic that posits violence as a possible tool for the creation of certain political forms and outcomes. In contemporary practices of political violence the organic metaphor – the notion that a body politic and its processes are akin to life processes and can be treated as such – is gaining currency, particularly as levels of sophistication in military technologies continue to improve. This is evident in current practice of using lethal drones for counter-terrorism operations, where drones are held to be a superior instrument to eliminate or reduce the risk of terrorism to the body politic. The former CIA director and chief architect of Barack Obama's drones programme, John Brennan, has repeatedly invoked the use of invasive violence for the health and well-being of the body politic and posits that it is indeed possible to cure the body politic of its terrorist cancer with the correct means. In a 2013 interview with the magazine *GQ*, Brennan invokes the 'cancer' metaphor several times, giving shape to the medical narrative with which violent political practices are justified (Brennan 2012a; 2013). Chapter 5 elaborates this medical metaphor as a justification narrative in greater detail. This type of biological narrative continues to be prevalent in US strategies in the war on terrorism today. By frequently associating the problem of terrorism to a cancer that must be contained at all cost, both the Obama administration and the Trump administration have perpetuated an image in which killing the unsavoury cells that have befallen a body politic is the best, the only option. The political goal is the expedient eradication of any dangerous affliction that befalls humanity – in all its unpredictable and ungovernable manifestations. The possibility of doing so, with minimal risk to those administering violence, is realised with new remote technologies such as drones or other robotic weapons systems operated from afar. The combination of the two facilitates a continual striving to eradicate all threats and risks, thereby potentially engaging in 'an endless series of wars to end all wars' (Brown 2003: 5). The entwinement of politics and violence is here expedited through new types of violent technologies.

Political history throughout modernity has been infused with violence as a political instrument and the Clausewitzian adage of war as an extension of politics by other means, and its various inversions,[1] have manifested themselves into the collective consciousness of political life for centuries. In 1962, Sheldon Wolin pointed out that theorisations of violence should take into consideration the respective values and foundations on which a community is based, because 'modern man's view of violence differs almost as greatly from pre-modern man's view as the nuclear warhead differs from the blunderbuss' (Wolin 2009: 41). Following his suggestion, a closer look at the interlacing of violence with political modernity, for our contemporary political forms, is in order if we are to understand the role of violence in techno-biopolitical politics today. What are the rationales and the values that overtly or tacitly underpin the role of violence in politics today? And how does this manifest in the relationship between violence and politics? As new processes and tools for violence emerge for our contemporary era, these questions arise anew. It has been suggested that investigations of political violence should take place not merely at the specific and particular level, referring to either particular ethnographic studies or specific types of violence, such as terrorism or militant civil strife, for example, but rather at the conceptual level as well (Keane 2004; Bufacchi 2007: 11). Particularly as new political and military practices emerge in the war against terrorism, which, to date, seem to have failed to quench the tide of violence that marks our contemporary era, the need to rethink the co-existence of the universal claim to humanity, and the notion of violence-as-politics in the name of life, is pressing. Where life has moved to the centre of political practices and imperatives, and violence is normatively conflated with politics as an instrument of power, two further questions must be tackled: what role does violence play in such a political society? And what norms does this produce for the relationship between politics and violence more broadly?[2]

In the following, I draw once more on Arendt for her analysis of political violence to spotlight the biopolitical rationales which inform justifications of violence in contemporary political practices. I begin this analysis by addressing her conception of violence as mute, instrumental and, importantly, problematic, if perceived as expedient political action. The chapter then moves to discuss the impasse that is produced when violence is understood as, or conflated with, politics; this has particular salience for a biopolitical context. Here, I want to draw attention to the instrumentality of violence when politics is (mis)construed as a form of management for specific ends. The notion of certainty for political outcomes constitutes, for Arendt, a misunderstanding at best and a delusion at worst. The chapter eventually ties the use of violence-as-politics to a *zoe*fied understanding of politics. These considerations form the basis from which to depart from an exposition of the core elements of an Arendtian understanding of life politics and its violent potential to then engage in an analysis of the ethical implications of political violence for the second part of this book.

Instrumental, impotent, mute violence

Conceptualisations of violence and politics in modernity frequently present the two concepts as inextricably, if not necessarily, linked, comprising a notion that the pragmatist John Dewey perhaps most succinctly and sombrely summed up as follows: 'nothing was ever accomplished without using force' (Dewey 2009: 8). While Dewey's sobering assessments may seem somewhat flippant,[3] it nonetheless echoes the modern sentiment that for certain political outcomes the use of violence is indeed justified, legitimate, if not necessary. The conflation of the two concepts is nothing particularly new: there is a wide-ranging body of scholarship on the interrelated conception of violence and politics, from a premodern Machiavellian tradition of understanding the use of force as integral to political rule, to the Hobbesian understanding of the essentially violent nature of humans which can be controlled by the very transference of this violence toward a mighty state structure; and eventually to theorisations of violence as a necessary, justifiable and perhaps legitimate means to disrupt forms of oppression in the pursuit of political freedom. Furthermore, in the comprehensive body of contemporary scholarship on collective violence, politics and violence are frequently presented together as indivisible concepts (Tilly 2003; Conteh-Morgan 2003; Barkan and Snowden 2000).

Throughout her writings, Arendt continually resists the conflation of politics with violence, although she was clearly aware that politics and violence have had a long tradition of enmeshment as 'no one engaged in thought about history and politics can remain unaware of the enormous role violence has played in human affairs' (1970: 8). In her seminal essay on the topic, *On Violence*, she offers a critical analytical engagement with 'violence in the political realm' and disentangles the two by making the case for violence as *antithetical* to politics and anti-political as such (1970: 35). The essay earned Arendt criticism for her radical distinctions of what does, and what does not, constitute violence, but it also secured her place as one of the most vital thinkers on the topic of politics and violence to this day (Bufacchi 2009: 3). In *On Violence*, Arendt does not offer a comprehensive definition of the term as such, yet it is clear in her writings that she distinguishes carefully between violence itself and the implements of violence. Where Arendt's analysis is most interesting for a conceptual understanding of the role of violence in politics, and also most contested, is in her positing of violence as distinct from the terms it is often (and in Arendt's view erroneously) conflated with, namely force, power or strength (Arendt 1970: 4). These distinctions have garnered much attention, as well as criticisms (Ricoeur 2006; Breen 2007; Hanssen 2000), but they make it possible to analyse violence as anti-political, as *outside* of politics, and vice versa. It is thus easier to grasp the concept of violence by paying attention to what it is *not*, in Arendt's account.

Violence, in Arendt's terms, is neither identical with strength (although phenomenologically it can be similar to strength) nor is it the same as force, which for her predominantly should be used to indicate 'forces of nature' or the 'forces of circumstances' (1970: 45), something that phenomenologically originates externally. She states most emphatically that violence is also not the same as power, or even authority. For Arendt, all these terms factor into a political context in modernity and become confused and conflated precisely when the dominant political question becomes: 'Who rules Whom?' (1970: 43). Arendt's theoretical conception of violence rests on some key claims about the nature of violence. Firstly, violence is always instrumental in nature. It relies on the use of implements and tools to amplify human strength. This means that, for Arendt, violence, in contrast to power, does not have an essence as such. Violence always requires 'guidance and justification through the ends it pursues' and thus 'cannot be in itself the essence of anything' (1970: 51). Secondly, violence and power are located at opposite ends of a spectrum (1970: 56). Although they often appear together, they are, for Arendt, distinct phenomena, whereby violence has the capacity to destroy power but cannot create it. This is a major point of contention for scholars commenting on Arendt's exposition on violence (see for example Ricoeur 2006) as it thoroughly defies commonly held perspectives that follow the Maoist position that 'power grows out of the barrel of a gun' (1970: 37). Arendt argues, however, that 'no government based exclusively on the means of violence has ever existed' (1970: 50). Any form of government intent on inflicting violence relies on others over whom it can assert influence, or power, for the execution of violence, especially on a large scale. Presciently, she notes that only with 'the development of robot soldiers, which ... would eliminate the human factor completely' would governance exclusively through violence become a possibility (1970: 50). Thirdly, and this aspect is closely related to the previous points, violence is located *outside* of politics and as such in itself anti-political. Violence, for Arendt, cannot be politics proper as it is itself 'incapable of speech' and as such inherently mute (Arendt 1998: 26; 2006a: 9). This, for Arendt, renders any modern conflation of violence-as-politics highly problematic.

That Arendt was acutely aware of the problems of modern politics and violence is evident when we consider her sceptical analysis of socio-political structures highlighted earlier. Here it is helpful to keep in mind her understanding of politics *proper*, as distinct from political administration or management as government. Recall that politics proper, for Arendt, is an action that takes place in public, among a plurality of participants, in their interaction for a common or shared interest (Arendt 2005: 93). It is this capacity for action and its concomitant capacity for new and unpredictable political outcomes that 'make man a political being' (Arendt 1970: 82). It is in the political coming together of people, who express themselves, their viewpoints and interests through language and speech, that political power is actualised. The mute character of violence excludes it from becoming manifest as politics proper. Moreover, where this possibility for political action is impeded, as it is the case in biopolitical structures of government, the role violence plays as a form of political expression appears to be magnified. Arendt identifies this ascendancy of violence over political action clearly: 'I am inclined to think that much of the present glorification of violence is caused by severe frustration of the faculty of action in the modern world' (1970: 83).

While Arendt was keenly aware of a *de facto* reality of political practices in which violence had been considered an integral part of politics for centuries, and specifically in her time – the twentieth century (Canovan 1995: 185) – she was nonetheless highly critical of the fact that the co-existence of politics with violence is, in fact, so ingrained in all modern generations that a disassociation of the two requires a conscious effort of imagination. The unquestioned concept of the construct of violence-as-politics thus presented a theoretical problem for her. It was precisely this lack of imagination, to conceptually and critically view this problematic composition in which modern politics has become imbricated with violent tools for the maintenance of government, and which Weber posed as the modern moral political dilemma for the 'courageous' politician, that Arendt addresses in her analyses of violence and politics in modernity throughout her work (Canovan 1995; Hanssen 2000: 25). While her perspectives on the problem of violence are most widely known through her expositions in the essay *On Violence*, theories of violence run through Arendt's entire body of work, not least as a central element in her analysis of totalitarianism in 1952. When this is read in the context of her critique of modern biopolitical societies as essentially anti-political societies, a clearer picture of the specific problems of this conflation of violence with politics in modernity emerges.

Violence-as-politics

While the violence-as-politics conjuncture in modernity posed a problem for Arendt, she was by no means a pacifist, despite her admiration for Gandhi's non-violent persistence (Canovan 1995: 187). In her analysis, she carefully considers the instrumentality of violence in situations of self-defence as potentially justifiable, and holds the violent struggle against tyranny and the closing of public spaces and spheres of political activity as not only a justified, and necessary, utilisation of violence but even as a legitimate struggle for liberation (Canovan 1995). Yet the idea of a linear means–end utilisation of violence as an expedient instrument to be employed for distinct political objectives is not an equation Arendt accepts, as she makes clear throughout her writings. Her account of violence is further problematised by the fact that she employs an instrumental justification for violence in some instances and simultaneously critiques the instrumentality of violence in the political context, specifically in the writings of Sorel, Fanon and Sartre. While she categorically rejects a means–end justification for violence as politics, she also makes concessions to possible justifications of instrumental (yet not political) violence, to achieve political goals. It thus seems that Arendt indeed creates pockets of justifiability for instrumental violent action for political goals, albeit in a very restrictive sense.

This makes the Arendtian account of violence as a political modus slightly fuzzy. However, a close reading of her account helps tease out a nuanced instrumentality of violence, one that considers violence as justifiable (albeit not legitimate) either when seeking to 'set the scales of justice right again' in non-political individual acts of violence[4] or, in a collective sense, in the pursuit of the 'liberation from oppression' and the 'constitution of freedom' (1970: 64; 2006a: 25). Arendt does indeed argue for violence as a means with which to gain certain freedoms, if – and this is the crucial point for Arendt – this violence as a means and a tool is able to provide the condition for a new beginning, a new body politic, a new constitution, the political ends of which cannot be entirely foreseen. It is the breaking open of politics *spaces* for which instrumental violence can be justified in the Arendtian account. This, as Christopher Finlay points out, moves closely along the lines of Walter Benjamin's conception of 'divine violence' (Finlay 2009: 40–1; Benjamin 2007: 297–300), in that it defends instrumental violence only for the *breaking up* of existing, oppressive structures, not – and this is an important difference in understanding Arendt's analysis of instrumental violence – to *bring about* a new structure (Finlay 2009: 40–1; Benjamin 2007: 297–300).

The perceived creative potential of violence is made manifest in a biopolitical modernity through metaphors that draw on imageries and processes of biology and nature, whereby power and violence are 'interpreted in biological terms', with 'life and life's alleged creativity [as] their common denominator, so that violence is justified on the ground of creativity' (Arendt 1970: 75). Therein encompassed lies another hazardous conflation for Arendt: violence understood as power. Where power, for Arendt, is 'indeed the essence of all government, ... violence is not' (1970: 51); where power has politically creative properties, violence surfaces where power is lost. Violence, however, cannot create power and is, as pointed out earlier, located at opposite ends in relation to power. Where power does not need justification, but rather legitimacy, violence, being instrumental in nature, relies on justification but cannot be (politically) legitimate (Arendt 1970: 52, 79). In other words, legitimation occurs in the political realm, from which violence is excluded. Violence can, however, be quite rationally justified by the effectiveness of reaching specific ends. This perceived creative force of violence is then also pivotal in understanding the justification of violence in a biopolitically informed political context and it is useful to briefly draw on Walter Benjamin's original essay *Critique of Violence* in order to elucidate the complex condition of political violence as always bound up with moral and ethical justification and legitimation narratives in the political context of a biopolitical modernity.

Benjamin's *Critique of Violence* presents a complex analysis of violence and its relationship with law, codifications and the state (Kochi 2009: 209). Similarly to Arendt's analysis, Benjamin is chiefly concerned with the means–end relationship of violence and law, whereby the essence of his critique focuses not on the ends that must justify the means, rather, any judgement of violence must take a more discriminating approach within the sphere of *means* rather than *ends* (Benjamin 2007: 277; Finlay 2009: 40). Derrida explains the basis of Benjamin's critique of justification in terms of the symbolic character in the very nature of violence as follows: 'the concept of violence belongs to the symbolic order of law, politics and morals ... And it is only to this extent that it can give rise to a critique' (Derrida 2002: 265). Where violence is differentiated from 'natural force' and placed in a political context, it is bound up with justifications and legitimisations. Thus, in constructing an approach that is echoed in Arendt, as highlighted earlier, Benjamin's focus of investigation is on the 'question of the *justification* of certain means that constitute violence'. Benjamin seeks to disentangle the justification of violence from natural and positive law as they both, in their antinomy, cannot shed light on the category 'violence' as such (Benjamin 2007: 279, emphasis added). In Benjamin's words: 'Natural law attempts, by the justness of the ends, to "justify" the means, positive law to "guarantee" the justness of the ends through the justifications of the means' (2007: 278).

Here, Benjamin arrives at a distinction between the implications of law and violence in what he considers as *mythical* violence and posits it against his concept of *divine* violence. Mythical violence, for Benjamin, is that violence which is both law-preserving and law-making. In Benjamin's analysis, mythical violence thus inevitably relates to existing state structures that prescribe, though laws, codes and norms, either a reaffirmation (preservation) of existing political structures or indeed the making of new laws and codes, thus also prescribing or reaffirming the normativity of practices in the context of political violence. In short, law-making violence has the potential to create new normative conceptions of existing political practices and powers, in an effort to overcome or modify prevailing power structures and create new conditions, new normative practices, new codes of law. Contained within this is a twofold function, as Benjamin writes:

> lawmaking pursues as its end, with violence as the means, what is to be established as law, but at this very moment of lawmaking, it specifically establishes as law not an end unalloyed by violence, but one necessarily and intimately bound to it, under the title of power. Lawmaking is power making and, to that extent, an immediate manifestation of violence. (2007: 295)

Contrary to Arendt's distinction between power and violence, it is also this violent entwinement of new law and power that then posits the spectre that underlies law-preserving violence, in which what has been established as law is to be preserved through legitimated violence or, at the very least, the possibility thereof. The Benjaminian mythical violence thus is concerned and intimately bound up with the making or preserving of law, of normativity, of new justifications of practices (2007: 297). The relationship between law and violence matters particularly within a biopolitical rationale, which seeks to ascertain as its legitimated and justified telos the life and progression of the population – an end that is at best always uncertain, and, in Arendtian terms, not a viable political goal. The biopolitical dimension of violence in the modern context was not lost on Benjamin. He specifically critiqued naturalistic conceptions of law, and the organic metaphor upon which the justification of violence for the attainment of something posited as natural is constructed, within the Darwinian justification that 'regards violence as the only original means, besides natural selection, appropriate to all the vital ends of nature' (2007: 278). For Benjamin, as for Arendt, it was clear that the 'dogma of natural history' serves to try not only to justify but to legitimate the violence that is used to attain 'natural ends' (2007: 278). Such an end to be attained through violence, however, is already outside the political realm in the biopolitical context. Furthermore, within the biopolitical rationale, the 'justness of the ends ... "justify" the means' in conceptions of natural law, while positive law ensures 'the justness of the ends through the justifications of the means' (2007: 278). Where Benjamin sees natural law and positive law as diametrically opposed, Arendt, with her insight into the development of the human-as-maker of life in modernity, allows us to understand how the two not only co-exist within the biopolitical rationale but serve as a more comprehensive and less politically contestable justification of political violence. The targeted killing programme with the use of drones, enacted by the US in the fight against terrorism is a case in point in this biopolitical rationale of justification as legitimation: while the survival and unhindered progress of humanity serves as the (natural) justified end, the technology used – the allegedly ethical, effective and minimally invasive drone

– serves as the justified means with which the justification of the ends can be ensured, and the practice thus legitimated. I will engage with the direct ethical implications of this practice in the next chapter.

It is in contrast to the identification of mythical violence with legal and normative violence that Benjamin takes a turn toward the messianic in his concept of divine violence. Benjamin's divine violence is a violence to end all violence; one that seeks to unravel the relationship between law and violence. Where 'mythical violence is law-making, divine violence is law-destroying'; able to establish a new order, one that is hitherto unknown, but of which it is clear that it will bring about a 'new historical age' (Benjamin 2007: 297; Manchev 2009). It is in this 'clean slate' approach that Finlay seeks to establish the parallels between Benjamin and Arendt.[5] It is important to stress, however, that these parallels are tenable only if Benjamin's conception of divine violence is not interpreted as creative, constitutive or in any way positive in terms of politics, for it is precisely the constitutive relationship of violence and politics (or with Benjamin: the law) that Arendt so vehemently rejects (Finlay 2009: 40). Violence as destruction presents an acceptable instrumentality in the political context for Arendt and in a messianic context for Benjamin. Violence as a creative force serving the notion of 'making' a new political history is, for Arendt, not possible and, for Benjamin, an impasse. Here we see that Arendt's critique of Fanon's alleged glorification of violence is most pertinent. An understanding of politics that is at once intrinsically bound up with violence and at the same time liberating from the perpetuation of violence is for Arendt problematic for the very fact that it has no creative capacity and thus cannot serve as a rational instrument for distinct political ends. Where Fanon, in his analysis of causal structures of violence, sees violence as constitutive of a new social order, he does so in the understanding that people can constitute themselves, effectively 'make' history come about, in the sense that Marx described when he so notoriously claimed that 'violence is the midwife of every old society pregnant with a new one' (Frazer and Hutchings 2008: 95). Even though Fanon sought to conceive of a new society that does not rely on violence, one that, in fact, is free of both the originating and liberating violence (Frazer and Hutchings 2008: 93), it is, indeed the immanent violence in the constitution of this new society as a distinct political end that makes this a dangerous perspective for Arendt.

Means–ends–outcomes: The certainty chimera

The difference here in Arendt's instrumental justification of violence and the violence-as-politics conflation is its relationship to the desired political *outcome*. For Arendt, uncertainty is immanent in all violent action. She writes: 'the danger of violence, even if it moves consciously within a non-extremist framework of short-term goals, will always be that the means overwhelm the end' (Arendt 1970: 80). In other words, violence, as enacted by humans among humans in the pursuit of political ends is subject to the same perils as all action is, namely that the outcome of human action cannot be certain. Certainty, for Arendt is a chimera in the violence-as-politics assumption. Arendt's most fervent criticism targets precisely this misconception perpetuated in modern political histories to date: the illusion of certainty present in the means–ends instrumentality for political goals and their modern glorifications. It is the appropriation of this original force of nature, violence, as a glorified means for a political end in which the aspiration of human deification that characterises modernity finds its most destructive extension.

The means–ends discussion surrounding violence, whether violence is violence as state-initiated violence, war or revolutionary violence, relies, Arendt argues, on the assumption that the ends can be controlled, if not ascertained. This is precarious for two reasons. Firstly, human action rests on plurality and difference, and always bears an immanent contingency so that the results of human action can thus never fully be controlled, not even by actors themselves, and, second, the very nature of violence contains within itself an uncontrollable, arbitrary element which is beyond the influence of actors (Owens 2007: 70). As Arendt states, '[n]owhere does Fortuna, good or ill luck, play a more fateful role in human affairs than on the battlefield, and this intrusion of the utterly unexpected does not disappear when people call it a "random event" or find it scientifically suspect' (Arendt 1970: 4).

Where violence serves its rightful instrumental purpose for Arendt is in the category work, in the act of fabrication of artefacts. As such, violence is intrinsic to the creation, to the making of a common *world*, not, however, in shaping political outcomes. *Homo faber*, the creator of the human artifice, has always been a destroyer of nature – here the instrumentality is limited and always for a certain, often expedient, end, whereby the end (the tangible work produced) justifies the violence inflicted (on nature) in order to gain the material for the object to be created. Thus 'the wood justifies killing the tree and the table justifies destroying the wood' (Arendt 1998: 153). In her assessment of the utility of violence Arendt does not see this violence employed for the creation of a world as problematic but rather as necessary for the construction of a shared worldly environment.[6] Typical of Arendt's strict categorisations and distinctions, the necessity of instrumental violence can apply only to the realm of work and within the very domain of *homo faber*. This delineation, however, becomes problematic when violence is conceived of as creatively instrumental in a modernity in which the boundaries between work and action have become somewhat blurred and labour, work and action become confused in a 'matter-of-course identification of fabrication with action' (1998: 306). If we consider that Arendt's aim was not to discredit the utility of violence entirely, but to seek ways that aimed at delimiting violence in the human realm, her strict distinction of violence as instrumental in fabrication, but never as politics, becomes clearer (Isaac 1992: 133). Her logic here is evident: where violence is limited to the making of a tangible, common world, it remains controllable; what is destroyed by humans can be rebuilt – to a certain extent. Where violence enters the realm of political action, it remains beyond control; what is destroyed by human violence cannot readily be rebuilt within intangible inter-human relationships.

A justification of instrumental violence thus cannot relate to the realm of action for the very reason that it has no creative, active capacity, only a destructive dimension. If destruction is an end towards creating space for something new, the instrumentality of violence is justified for Arendt. If violence is used to bring about a political beginning, it will fail, as such interpersonal outcomes cannot be secured through violence. Uncertainty is implicit in politics proper for Arendt. In an Arendtian biopolitical modernity the confusion of 'making' and 'acting', and consequently the conflation of violence as a means for political ends, is further complicated as it subsumes ideas of political theory at the time. Again, here she takes issue with the notion that violence can be instrumentalised to 'make' history and, rather than constituting an always precarious element in the broader tapestry of political events, humans become makers of their own destiny, self-constituted in their political realisation. Once such thinking reigns supreme, risk and uncertainty are unacceptable and everything that derails a specific telos for the determined destiny of humanity becomes an obstacle to be overcome by all means. The means then justify the end – a dangerous foundation on which to stipulate the coming about of political communities, because it may always prioritise violence for its ends. Adding qualifications to limit the types of means are, for Arendt, not enough because they 'take for granted a moral system which … can hardly be taken for granted, or they are overpowered by the very language and analogies they use' (1998: 229). Implicit in the logic of ends always already is the justification of means, and more so still when the metaphors used refer to that most foundational end: the biological health and survival of a body politic.

When humans see themselves, politically, as the makers of their political trajectory, in a direct instrumental way, violence, as the handmaiden for a new political history to come is not limited but rather exacerbated. In Arendt's words:

> Only the modern age's conviction that man can know only what he makes, that his allegedly higher capacities depend upon making and that he therefore is primarily *homo faber* and not an animal rational, brought forth a much older implication of violence inherent in all interpretations of the realm of human affairs. (1998: 229)

This modern understanding of humans as 'makers' of history is for Arendt doubly problematic. Not only does such a means–ends theory of politics instrumentalise political action but it also establishes that political ends stand under the mandate of purpose, efficiency and expediency in modernity. In other words, a level of rational certainty is introduced into the realm of modern politics that accepts all means as long as they are *efficient* in achieving a desired end – the securitisation of society. This, for Arendt, is mechanised through bureaucratic structures that pose as political technologies. And precisely therein lies the enormous potential for violence for Arendt:

> the greater the bureaucratization of public life, the greater will be the attraction of violence. In a fully developed bureaucracy there is nobody left with whom one can argue, to whom power can be exerted. Bureaucracy is the form of government in which everybody is deprived of political freedom, of the power to act. (1970: 81)

In other words, biopolitically informed bureaucratisation marks the modern social realm and renders public life anti-political. Julian Reid frames this development in contemporary biopolitics similarly, in terms of 'logistical' life, which suggests that life today is under duress, forced to be logistical and efficient (Reid 2006: 17–39). This, however, does not only apply to the political administration of populations, but extends to the mathematically and statistically ascertainable individual as well. Where mapping of life processes and human behaviour become possible, human development and action can be rendered in operational terms, whereby the 'oldest conviction of *homo faber* – that "man is the measure of all things" – advanced to the rank of a universally accepted commonplace' (Arendt 1998: 306).

As we link this back to the previous chapters, it is evident that Arendt drew a clear and critical link between this mathematisation and analysability of the human and the technologisation of life in modernity. In Reid's work we see the primacy of purpose, manifested in logistical structures, that seeks to legitimise violence in modernity. He states: '[t]oday it is deemed necessary to defend the logistical life of society from enemies which threaten to undermine the efficiency of society itself' (2006: 35). The problematic character of the rise of society, where life-processes and the notion of necessity become paramount, is here highlighted once more. It is in the biopolitical rationale that violence

> can indeed be easily understood as a function or a surface phenomenon of an underlying and overruling necessity, but necessity, which we invariably carry within us in the very existence of our bodies and their needs, can never be simply reduced to and completely absorbed by violence and violation. (Arendt 2006a: 54–55)

In a rational, technologised modernity, the will to certainty and control, and the compulsion of necessity and purpose, are both dependent on, and simultaneously constitutive of, violence as an instrument for political achievements. Yet, in Arendt's view, it is precisely this violence as a means for an allegedly specific political end that can only ever reconstitute the violent assumptions in modern politics. In the totalitarian context, this aspiration of omnipotence, of not only fabricating artefacts but also making, or *bringing about* humanity, was reflected in the total integration of 'man as a whole' (Arendt 2004: 445) – private and public, body and mind – within the totalitarian machine and its genocidal consequences. In the Cold War period, the aspiration to this quasi-divine omnipotence was made manifest through the development of technological weapons capacities as a means that by far usurped the very ends for which they were created, thereby rendering the means–ends category in modern politics grossly absurd.

The relationship between such instrumentally understood violence and political ends carries for Arendt a comprehensive problem: 'As long as we believe that we deal with ends and means in the political realm, we shall not be able to prevent anybody's using *all* means to pursue recognized ends' (1998: 229, emphasis added). Her concerns must be read clearly against a background of the rationalised violent means for political ends in the totalitarian regime of Nazi Germany, yet, given the lasting structures in contemporary modernity, this continues to have relevance. When we consider Arendt's account of life politics, it becomes clear where the dangers of such instrumental thinking in politics lie. As life processes become absorbed into the political process, and matters of biology are perceived of as unfolding on a historical continuum in a Darwinian insistence of natural movement as 'not circular but unilinear' (Arendt 2004: 597), certain actors as the self-regarded makers of said history, and thus humanity, can only but be tempted to use *all* means to bring history into being, including 'creative' violence. Zygmunt Bauman convincingly follows this trajectory for a characterisation of genocidal violence in modernity.

In a similar vein as Arendt's critique on the Marxist idea of 'making' history come into being, Bauman sees in the very utility of violence a quintessentially modern perspective that rests 'on a devious symmetry of assumed intentions and actions' (Bauman 2012: 91). What is so distinctly modern in this relationship is the very idea of purpose in the application of violence towards a political end, whereby the end in this context is no longer the elimination of an enemy or an adversary but rather the making of a society, one that is 'objectively better', in line with superior values and in accord with an 'overall, scientifically conceived plan' (Bauman 2012: 91). It is then the very ideology of an objectively superior society that allows for the emergence of what Bauman likens to a 'garden culture', which departs from the underlying assumption that it is possible to create an ideal human world – 'more efficient, more moral, more beautiful' (2012: 92). Such genocidal vision not only relies on the eradication of 'contingency and chance' (2012: 90) but also could not exist without the practice of 'scientific management', specifically medical science and metaphors (2012: 73). This point gains further salience in light of the rapidly developing scope and scale of the technologisation of political violence in contemporary modernity, as I will discuss in later chapters.

In 'Introduction into Politics' (Arendt 2005) Arendt presents perhaps the clearest account of why she is so adamant in rejecting a conflation of violence with politics. In this meticulous essay on the meaning of politics in contemporary society, she elaborates the fragile relationship of means and ends in politics in the relation to violence. It is here that her account of the means–ends construct of modern political violence is most nuanced as she discusses the problem of war and revolution in modern politics. Crucially, Arendt draws a clear differentiation between political *goals* and political *ends*. While goals are indeed what political action pursues, they can only serve as a directive or guideline for the respective political action (Arendt 2005: 193). In other words, in political action, the outcome can be aimed at but never be ascertained. This uncertainty, the contingent, is immanent to politics proper for Arendt as it relies on action by humans and among humans, which has no clear and immovable outcome, but remains dynamic. 'Concrete realisations' in politics 'are constantly changing because we are dealing with other people who also have goals' (2005: 193). It is only when the use of force – violence – comes into play as political action that these changeable goals have the significance of ends in a quasi-rational and scientific process. Violence as political action, then, seeks to attain an end, for it is only the end that can ultimately justify the violent means.

But what does this distinction between political goals and political ends ultimately mean for Arendt? It suggests two crucial aspects in the relationship of politics and violence: political goals become political ends when politics is misunderstood as 'making' and political action is expressed through violent actions. It is precisely in this distinction that Arendt excludes the violence-as-politics construct. In other words, the pursuit of political goals, which are never fully realised is, in the Arendtian understanding of politics, conducted through the 'back-and-forth' of speech across the interests and spaces between people. When violence is introduced, political goals become ends, for violence stands under the mandate of expediency in its immanent instrumental character. When political ends are confused with goals and political violence is seen as political action, where 'nothing counts except the achievement of postulated and fixed ends, brute force will always play a role' (2005: 194). In this confusion of political violence with political action, a particular rationality is ascribed to such violence. Arendt's position on violence-as-politics becomes clearer when read against some of her other texts. As she writes in *On Violence*:

> Violence, being instrumental in nature, is rational to the extent that it is effective in reaching the end that must justify it. And since when we act we never know with any certainty the eventual consequence of what we are doing, violence can remain rational only if it pursues short-term goals. Violence does not promote causes, neither history nor revolution, neither progress nor reaction; but it can serve to dramatize grievances and bring them to public attention. (1970: 79)

While she maintains that the act of violence is an anti-political act, as it is mute in itself, she makes concessions as to its ability to dramatise *political* issues, albeit in a limited fashion, as she immediately warns: 'if goals are not achieved rapidly, the result will be not merely defeat but the introduction of the practice of violence in the whole body politic' (1970: 80). The danger is thus in the confusion of political action with violent acts: when mute violence becomes modern politics' form of expression through mute actions that are rationalised and made clinical, the assumption that it is controllable as a means for a political end is radicalised. It fully leaves aside contingency.

The inherent muteness of violence presents a challenge to its study as a political factor for Arendt. It is precisely for the lack of speech, or language, in violent acts that political theory 'has little to say about the phenomenon of violence' (Arendt 2006a: 9). Consequently any theory of war or of revolutions must deal with the *justification* of violence, whereby the justification is constitutive of its political restrictions. It is precisely when this distinction is no longer observed and maintained that theories of wars and revolutions 'arrive at a glorification or justification of violence *as such* [and] is no longer political but anti-political' (2006a: 9, emphasis added). It is true that, for Arendt, violence in a purely mean–ends understanding of the phenomenon, presents a much more immediately effective instrument to achieve a defined and declared end, so much so that, if we are speaking in purely utilitarian terms, violence can be seen as a much more effective substitute for action in the pursuit of a political aim. In Arendt's words: '[i]f nothing more were at stake here than to use action as a means to an end, it is obvious that the same end could be much more easily attained in mute violence, so that action seems a not very efficient substitute for violence' (Arendt 1998: 179). Here again we could be tempted to understand Arendt as conceding that violence can indeed pursue an end, perhaps even quite effectively. But the key here is the emphasis on utility, which for Arendt is not part of the realm of politics and it is important to understand in this context Arendt's assertion of action as explicitly refusing the means–ends category. Recall that action, unlike violence, is not instrumental and not solely utility-oriented, but rather facilitates new beginnings with often uncertain outcomes – something that violence is utterly incapable of. Action, for Arendt, is thus the essence of politics, not violence. But in this construct in modernity, where difference, plurality, freedom and togetherness based on interest and a shared world becomes diminished – in short a reduction of the possibility for politics – lies a distinct danger as action is absorbed by behaviour, into a subsuming means–ends structure:

> This happens whenever human togetherness is lost, that is, when people are only for or against other people, as for instance in modern warfare, where men go into action and use means of violence in order to achieve certain objectives for their own side and against the enemy. (Arendt 1998: 180)

When this happens, when utility becomes the metanarrative of society, violence can indeed be seen as the more efficient and effective means to achieve a political end. The fallacy of this assumption is clear but bears repeating: the supposition that violence as action could achieve distinct and clear goals is a chimera. And in fact, the further the means are from the desired goals, the less likely it is that violence can achieve anything politically.

The anti-political frustration of action in a biopolitical context enables the use of violence in a second way: as the only form of expression by those otherwise muted and rendered anti-political. The Arendtian category of action as the only performative expression of politics proper, and its relationship to violence, plays a key role in this. In a modern polity that is dominated by life politics and thus by the very restrictions of necessity, freedom, the very basis for political action proper, as Arendt understands it, is hampered. Here again we must remember Arendt's critique of the modern glorification of violence as a reaction to a significant impediment to the capacity for action in modernity (Arendt 1970: 83). Where politics proper is stifled, as action is stifled, violence, otherwise mute, becomes the most radical dramatisation of the expression of individuals and groups in a political society – it serves as a demonstration, not an explanation of contestation. In other words, the frustration of the faculty of action in modernity renders violence a performative means of expression, often misunderstood as power. Arendt explains this as follows: 'it is simply true that riots in the ghettos and rebellions on the campuses make "people feel they are acting together in a way they rarely can"' (1970: 84). In other words, when language and speech become muted in an anti-political society, violence, as perceived action, replaces expression, often without a distinct goal other than the instrumental destruction in reaction to existing, frustrating structures. And it is in this life-political constellation that her warning resonates most clearly: '[t]he practice of violence, like all action, changes the world, but the most probable change is to a more violent world' (1970: 80).

Conclusion

In a social and political environment where the physical functioning of the individual human and the life, survival and progression of humanity stand fundamentally at the heart of political aspirations, the use of violence for political ends becomes a complex and problematic constellation. Prominent conceptions of politics that uniformly confirm the presence of violence where politics appears have fostered the view that violence may indeed constitute the very essence of politics (Bufacchi 2007: 190)[7] and in a modern era that is characterised by large-scale, total warfare, vast numbers of civilian casualties, genocidal practices aided by ever – more murderous technologies, we are increasingly inclined to accept as truth that 'there could be no politics without violence' (Duarte 2006: 408). While statistics gathered by the Human Security Report Project highlight that occurrences of interstate war have decreased in the past six decades,[8] a decrease often related to growing levels of democratisation of governments around the globe (Doyle 1983; Keane 2004: 2–5), there is ample evidence, for example provided by the Stockholm International Peace Research Institute (SIPRI), that forms of one-sided violence, directed toward civilians, have, in fact, risen. In particular, the changing nature of contemporary warfare seems to shift the lethal risk from combatants to civilians. In light of a growing number of theatres of military engagement in the past decades, paired with the ever growing intrusive and destructive capacity of weapons technologies, the perceived and actual rise in the scale and scope of violence across the globe supports the assertion of a number of scholars that modernity, until now, constitutes an age of violence (Hobsbawm 2007: 141; Bufacchi 2007; Campbell and Dillon 1993). This sentiment is echoed by others who highlight the fact that the spread of democracy, despite all hopeful proclamations, has not been successful in limiting levels of violence; in fact, some would argue, quite the contrary is the case (see for example Worcester, Bermanzohn and Ungar 2002). Indeed, the argument has been made that both levels and forms of violence – and in particular political violence – seem insufferably high in recent times (Bufacchi 2007: 1; Kreide 2009) and pervasive throughout the entire global realm (Jabri 2007: 1). A brief glance at the data from nearly two decades of the war on terrorism seems to indicate that violence shows no signs of abating as the mute dialectic between suicide terrorism and lethal drone strikes continues to destroy a growing number of civilian lives.

The means–ends logic of 'you can't make an omelette without breaking eggs' (Arendt 1998: 229) has indeed become so prevalent that it seems difficult to see how it could be otherwise, or how this conflation might in fact produce counterproductive outcomes. The logic and processes implicit in biopolitics foster not only a limitation of politics proper by requiring conformity and bureaucratising political processes, but also carry within them an immanent moral mandate. The health and survival of the population is the ultimate, the prime political good, for which violence is a justifiable tool in ensuring this goal. This 'paralyses any ethics and any politics', as Agamben notes (2000: 116). Aided in this process of means–ends conflation and confusion is the alleged neutrality of new military technologies. The use of drones in the relatively recently instituted practice of targeted killing is exemplary for the ethical framing and neutralising of violence as a creative political tool. Chapter 7 devotes its attention to this in more detail. In the next chapter I depart from an Arendtian account in discussing how ethical rationales prevalent in contemporary politics – national and international – are equally infused with biopolitical perspectives and aid in providing a justification framework for acts of political violence.

Notes

1 Both Arendt and Foucault turned the dictum on to its head to come to an understanding of politics as the continuation of war by other means, although each of them in a different context.
2 A brief word on defining violence. The term violence notoriously eludes definition, as the specificities of actual violence are manifold and theorisations of the

term are often polarising and steeped in concepts of morality. Aside from the normative question whether violence can or should ever be framed in terms of 'good' violence in opposition to 'bad' violence (Mueller 2002), abstractions and realities of violence are inherently elastic. Conceptions of political violence in contemporary discourses are often addressed in terms of structural violence in contrast to direct violence, most influentially outlined by Tord Høivik and Johan Galtung (Høivik and Galtung 1971), and have been largely classified in terms of their breadth or narrowness. The problem remains, however, that in order to discuss a conceptual understanding of violence one must find a definition that is broad enough to not unnecessarily limit certain aspects of violence yet 'narrowly enough to be useful' (Litke 2009: 297). C.A.J. Coady has effectively shown that different approaches (structuralist, legitimist) have political significance and warrant careful consideration when attempting to define violence as a concept (Coady 2009: 244). Consensus as to a universal definition of the term, however, does not truly exist among scholars, and finding a definition is complicated further by the very 'evolution' of the application of the term (Keane 2004: 30). The term political violence presents a sub-category of a far-reaching general concept of violence (Bufacchi 2007: 96), and, while this narrows the scope considerably in attempting to find a useful definition for analysis, there are still wide-ranging perspectives as to what should be included or excluded by the term. Definitions are often normatively tinged in framing violence in terms of legitimacy or legality in a political context. Such normative definitions are often restrictive and can, as such, prove counterproductive in the examination of violence as a concept. A more neutral definition of violence offers a broader platform from which to begin any analysis (Keane 2004: 35). But 'normative flourish' is not the only obstacle in finding an appropriate definition; there is also a wide elasticity in what is considered violence in a political context. Where some scholars focus their investigations of political violence on the notion of the unwanted violation of physical integrity or physical harm inflicted in a political context (Keane 2004; Rapaport and Weinberg 2001), others broaden the definition to include psychological or mental harm as inflicted by a deliberate omission to act (see for example Litke 2009; Bufacchi 2007; Bäck 2009). And, in addition to the physical and psychological aspects of violence, the subject might further be considered in its personal or structural manifestations, as Johan Galtung carefully delineates in his influential essay 'Violence, Peace and Peace Research' (Galtung 2009). Other, broad concepts of violence in a social context include occurrences of alienation (i.e. living conditions) and social exclusions, as well as instances of repression such as the divestment of social or political rights (Salmi 2009: 311–18). The aim of such definitions of violence is to address the far-reaching forms in which individual persons and groups can be oppressed and harmed in a political context. What makes such broad definitions problematic, however, is that they render political violence so general that they become all-pervasive in society and a meaningful analysis of violence as a political concept becomes further complicated, while that which is particular to violence as it is conventionally understood, i.e. the use of physical force to cause harm to the integrity of the subject of violence, becomes somewhat obscured (Giddens 2002: 246). As the subject of my analysis in this discussion relates primarily to an Arendtian conception of violence, it makes sense to keep to a more narrow and conventional understanding of violence in this context. In Arendt's writings, violence is predominantly understood in its physical phenomenology, as an instrument whose very substance is 'ruled by the means-end category' (Arendt 1970: 4) and less so in its structural manifestations. Despite this circumscribed understanding of violence in a more narrow sense, however, her conceptual account of political violence is by no means uncomplicated and warrants a more nuanced analysis.

3 Specifically Dewey's claim that 'squeamishness about force is the mark not of idealistic but of moonstruck morals' presents a highly normative perspective in this discussion but represents an existing polarisation between advocates for the elimination of violence as politics (Dewey 2009: 8).

4 Arendt elaborates this through an interpretation of Melville's Billy Budd, whereby Budd represents for her a classic example for acts of violence to gain 'justice'. However, this 'justice' is not political justice for Arendt but, owing to its emotional content, anti-political (Arendt 1970: 64; Owens 2007: 99).

5 Both Walter Benjamin and Hannah Arendt have produced some of the most frequently cited works on the concepts of violence, both were contemporaries in the horrors of the politics of National Socialism in Germany and its most radical violent manifestation, both were highly sceptical critics of modernity and they were friends. Where they make an interesting pairing against a shared background of a totalitarian experience of terror, it is surprising that there is very little work that sheds light on the relationship between Arendt and Benjamin, specifically with respect to their work on violence. However, what is perhaps even more surprising is that Arendt, who was such an astute scholar of violence and was a driving force in publishing some of Benjamin's work posthumously, makes, throughout her work, no mention at all of Benjamin's exposition on violence (Finlay 2009).

6 In her ready acceptance of violence inflicted on nature for the construction of a common world for humans, it seems that Arendt shows an implicit acceptance of the domination of humans over nature as she readily accepts that natural objects may, unproblematically, be turned into objects that serve humans and thereby indeed become reified in their value. New materialist accounts may challenge such a perspective.

7 While Bufacchi, in his 2007 investigations of violence and social justice frames his thesis around the premise that violence constitutes the essence of politics in his introduction, he later, in his concluding chapter, questions this fact and reasons eventually that it can, in fact, be the essence of violence (Bufacchi 2007: 5, 187–97).

8 According to the Human Security Report Project, interstate violence in fact peaked in the late 1980s and only began to fall again in the past two decades. This is often attributed to greater levels of democratisation across the globe

– the quintessential argument of the democratic peace theory. This, however, remains contested (see for example Singer and Small 1976; Spiro 1994). Whether the levels of overall occurrences of violence in a political context have risen or fallen in recent decades is far from being settled. Research from the Uppsala Conflict Data Programme indicates that violent conflicts have actually decreased in number, but other studies show that the campaigns of violence have significantly increased since the 1990s (Stepanova 2009).

5

Ethics as technics

Without silence, without the hiatus, which is not the absence of rules, but the necessity of a leap at the moment of ethical, political or individual decision, we could simply unfold knowledge into a program or course of action. Nothing could make us more irresponsible; nothing could be more totalitarian.

Jacques Derrida, *Adieu* (1999: 117)

The biopolitically informed relationship between politics and violence is, as the previous chapter has shown, infused with problems. The relationship between ethics and politics is equally complex. Arendt controversially engaged with this relationship by drawing a sharp distinction between morality and politics, arguing that morality is not only separate from politics proper but highly personal, as an essentially private matter; thus, she suggests, it ought to be excluded from the political (public) realm (Canovan 1995: 155–7; Kateb 1984: 29; Owens 2007: 107). Arendt's grappling with the complexities of morality and politics reflects her own experience with the collapse of morality in Nazi Germany, and her struggle to make sense of the problematic and precarious relationship between morality and politics in a modernity that had demonstrated a hitherto unseen appetite for destruction and annihilation (Arendt 2003a: 54; Canovan 1995: 156; Meade 1997: 109–10). In an effort typical for Arendt, she sought to understand what happens to morality in the public sphere when it becomes unanchored from a fixed, supra-human law-giving authority and arrived at the insight that ethics, like politics, is never fixed but rather has to be negotiated and ascertained ever anew in agreements that may be enshrined as norms, customs, rules and standards for a time being (Arendt 2003a: 50; Kohn 2003: xviii).

Acting politically, and thinking and judging are two crucial elements in Arendt's considerations of morality and moral failure – an engagement reflected particularly in her writings on the Eichmann Trial in 1963. It was also her controversial coverage of the trial and the subsequently much debated and discussed idea of the 'banality of evil' that informed her philosophical thinking about morality and judgement in her final work, *Life of the Mind*. This engagement, however, remained incomplete. Nonetheless, despite the fact that her exploration of 'the basic question of ethics' remained unfinished (Ludz 2007: 797–810), Arendt certainly touched upon an important aspect of considering morality and politics in modernity.[1] However, the aim of this book is to examine the biopolitical underpinnings of ethical considerations relating to the perplexities of our own time. Arendt's unfinished theoretical insights into ethics and morality in modernity are limited for this aim. It is thus here that I depart from an Arendtian exposition, and engage the biopolitics framework identified earlier to examine the ethics of political violence at work today.

The ethics/politics nexus

Against a contemporary backdrop, where hitherto morally prohibited practices such as the extrajudicial targeted killing of individuals[2] are permitted as legitimate, justifiable and even necessary practices by a growing number of countries, it is prudent to consider the relationship between politics, violence and ethics – and its limits – anew; even more so, when such practices are underwritten by technologies that clearly shift the horizons of ethical thought and practice. Questions of good and bad, right and wrong are inseparable from basic political concerns about legal, procedural and institutional organisation in a political community (Hutchings 2010: 8). Especially since 9/11, and in the ongoing war against terrorism, are practices of political violence infused with moral language and concerns. The mandate for security, which is engendered by a life-centric politics, bears within it a certain moral righteousness already. This does not, however, solve the problem of ethics of violence, rather it gives form to justifications of 'productive' violence in the securing of life. The notion that violent political acts ought to be considered in terms of right or wrong, good or bad and all shades in between has been a recurring demand in political theory and practice, captured in the various discourses on the just war tradition, which dominates the subject of the ethics of international violence to this day (Brown 1992: 132; Rengger 2013: 7–8).[3] The inherent open-endedness of the question of ethics in political violence is perhaps unsurprising when we consider the complex and dynamic nature of the international sphere. And indeed, specifically in an international context, matters of violence as an instrument of foreign policy tend to lack an in-depth engagement with the question of ethics, reiterating the stale division between morality and the practical realities of war and international politics all too often (Bulley 2013; Shue 1995; Rodin and Sorabji 2006).

At the heart of this might be the still ill-demarcated and somewhat underexplored relationship between ethics and politics (Chadwick and Schroeder 2002: 15). In this, ethics in relation to politics, specifically international politics, is predominantly understood in practical terms, whereby ethical concerns are considered as a related, but separate dimension to politics. A recent proliferation in scholarship on global ethics notwithstanding,[4] ethical concerns addressed in the international and global sphere are primarily framed in terms of finding and applying appropriate ethical principles, codes and rules in trying to resolve 'real moral problems' (Coombs and Winkler 1993: 2). Such a perspective is of limited scope for the complexities of a plural international domain and inevitably produces a tension that privileges practical concerns for ethics. Yet, far from being settled, the general sub-field of applied, or practical, ethics remains contested in methodology and approach (Beauchamp 2005: 7–14; Chadwick and Schroeder 2002: 1).

The previous chapter has addressed the precarious conflation of politics and violence in modern biopolitics and has sought to point toward some of the dangers that Arendt identified as a symptom of this specifically modern variant of the understanding of violence-as-politics. It is clear that biopolitical practices open up scope for potentially new or altered moral dimensions that require addressing. What are the fundamental moral principles substantiating norms and normalisations of ethical practices of (bio)political violence today? What do these moral principles rest on? And what renders such moral principles and ethical practices acceptable in a life-centric political society (Hutchings 2010: 9)? In order to investigate the relationship between biopolitical practices of political violence and their relationship to ethics in contemporary modernity more closely it is helpful to take a brief look at the foundations upon which these types of ethics rest themselves. To do so an engagement with the field of applied ethics and its contemporary relevance for politics is required.

In this chapter I argue that contemporary conceptions of political ethics, as a sub-field of applied ethics, are infused with biopolitical rationales, and enable the justification of acts of violence through a biopolitically informed lens. Metaphors drawn from modern professionalism as well as the medical field, are instrumental in this, as they produce, I argue, an *adiaphorisation* – or neutralisation – of violent practices, on one hand, and a radical coding of ethics in violence on the other. Both strands of the modern biopolitical rationale – the politicisation of *zoe* and the *zoe*fication of politics – are reflected in contemporary ethical justifications of political violence in modernity. I suggest here that the techno-biopolitical rendering of life shapes a perspective of ethics as code, rule and law, while the *zoe*fication of politics supports narratives that seek to justify the use of violence as necessary for the continuation and progress of humanity. This is particularly clearly reflected in the medical metaphors used in contemporary counter-terrorism interventions. The chapter first considers the wider context and critiques of practical, or applied, ethics before addressing the biopolitical rationales and limits of thinking about issues in international politics under the rubric of applied ethics. I then trace, in the final section, the medical narrative, which is increasingly deployed today for justifications of violent interventions in international politics, and highlight how the language and frameworks used impede the possibility for ethical contestation.

Deus ex homini

Today, as we find ourselves within a radical scientific-technological sociopolitical ecology in which science and technology determine knowledge and subjectivities more than ever, conceptions of ethics are in a process of redefinition. No longer closely tethered to a transcendental authority for the provision of both anchor and architecture for ethical frames of reference, ethical approaches today reside within varied doxas and precepts. With a shifting understanding of the human not merely as a 'given' entity of divine creation but rather as an actor who generates both history and life, the possibility of certainty about right and wrong becomes precarious. In such a sociopolitical condition, the biopolitical developments in technology and science replace transcendental divinity as the ultimate source for determining and instructing 'right' and 'wrong'; in short, the human, as maker of both life and history, now commandeers this space. Amy Swiffen identifies in this shift a biopolitical relation. She notes:

> It is true that the idea of God as a supreme lawgiver has given way to the notion of the authority of nature defined wholly by reason and scientific knowledge, not metaphysics. Along with this comes the belief that moral principles can be directly known by human beings. (2011: 39)

Empowered by scientific-technological advancements that produce new insights and knowledge about nature, the human and life itself, 'forms of ethical thought that are based on limits to human knowledge and power are no longer justified' (Swiffen 2011: 64). For Swiffen, this yields a new moral mandate: to shape and secure the life of future generations in a biopolitical turn. What emerges consequently is a 'new natural law', which replaces theological provenance with the rationality of natural processes, and the 'certainty of moral knowledge which asserts that survival is the minimum purpose of life, which is seen as continuous with nature' (2011: 92). The relationship to law in this biopolitical rationale of ethics is essential here as it emphasises the relevance of juridical structures for the practicality of ethics in modernity. It is in this turn to natural processes for ethical knowledge, and the format of the law, that ethics becomes something to be pinned down.

Despite the shift in authority for moral principles, away from a divine entity to the human as such, the notion of ethics as tethered to some sort of 'law-giving' authority remains prominent in discussions of ethics in politics within a secularised modernity. This is particularly relevant for the sub-field of applied ethics (Bauman 2000: 85–6). Law, here, works in two ways: it can, as Tom Beauchamp lays out in his text on the nature of applied ethics, be considered the public agency for making morality intelligible as social guidelines, whereby case law has been substantially influential in all areas of applied ethics (Beauchamp 2005: 2). Furthermore, as Bauman notes, in modernity, ethics is shaped predominantly after the *format* of law, so as to structure moral content as ethical code (Bauman 1993: 29). Arendt made similar observations when she reflected on the relationship between ethics and the guidelines that give shape to ethical content. Expanding on her argument that moral rules and standards can be changed 'like a table cloth' (Arendt 2003a: 50) in modernity, she recognises that

> what people get used to is less the content of the rules ... than the possession of rules under which to subsume particulars. If someone appears who, for whatever purpose wishes to abolish the old 'values' or virtues, he will find that easy enough, provided he offers a new code. (Arendt cited in Meade 1997: 123)

Where the rule form of ethics is paramount, the danger of a conflation of ethics with legislative regulations in order to secure 'good' behaviour looms large. Furthermore, the growing recognition of legal frameworks as a source of international authority in deciding matters of war and violence has propelled considerations of moral content into somewhat of a grey area (McMahan 2008: 19–20). Contemporary discussions on the ethics of drones, for example, are testament to the conflation of thinking ethically about lethal drones with the requirements of adhering to existing laws for the use of drones. The focus on law as primary reason, superseding individual ethical judgement and more importantly replacing the potential indeterminability of ethicality with a rational ethico-legal framework, betrays a certain rationalist perspective in turning to practices and procedures that, mirroring the scientific approach, seek to validate moral judgement and gives rise to the prevalence of an ethical rationalism that holds the potential to become a moralising force in modern society. As Bauman explains, '[l]aw and interest displace and replace gratuity and the sanctionlessness of moral drive. Actors are challenged to justify their conduct by reason as defined either by the goal or by the rules of behavior' (2012: 214). He recognises the relative content of morality as socially normed: 'The moral authority of society is self-provable to the point of tautology in so far as all conduct not conforming to the societally sanctioned rulings is by definition immoral' (2012: 213). The regulation of behaviour through normed ethico-legal codes and frames becomes paramount. In order to make ethics a regulatory theory, the homogenisation of humans and their conduct is crucial as informing a norm of behaviour. The application of ethics through regulatory frameworks, guidelines and codes has thus a functional dimension.

Ethics itself then is turned into something that must serve a purpose or, at the least, have an underlying reason. This finds a radical expression in discourses on the evolutionary purpose of ethics in the quest to find what ethics is actually 'good for'. Biological determinism discourses most starkly exemplify the aim to mitigate the indeterminability of ethics and ascertain the functionality, if not performativity, and 'success' of ethics, as well as society's ability to predict certain outcomes of actions, especially in instances of moral ambiguity through an investigation of the 'biological roots of moral behaviour' (Wilson 1998). In his influential work on biological science, Edward O. Wilson went as far as to suggest that, in fact, the inquiry into ethics ought to be removed as a study of philosophy and become a 'branch of biological science' (Rodd 1990: 84). This move would ground ethics in a 'foundation of verifiable knowledge of human nature sufficient to produce cause-and-effect predictions' (Wilson 1998). Wilson's pursuit to ground ethics in biological foundations, so as to situate it as verifiable, if not predictable, is, perhaps, the extreme manifestation of the problematic of ethics in a biopolitical modernity, but, with its focus on the biological underpinnings as a determinant of human behaviour, it epitomises the desire to render the human and her actions calculable and ascertainable in the search for certainty and predictability. Contemporary research in neuroscience picks up on this strand of inquiry, in a new biological determinism turn, when it targets the notion of free will as merely a chain of synaptic sequences. In other words, according to such neurological studies, a thief's decision to steal is determined by their neurological make-up and conditioning, by the unconscious brain which makes the decision for them. Suggested here is a limited moral agency, whereby 'immoral behavior is not necessarily the product of willful acts' (Tancredi 2005: 9). In these terms, free will is an illusion and casts a serious doubt over the ethical category of responsibility as such.[5]

In this particular rationale, the biopolitical underpinnings are clear. Ethics resides within neurological structures that can be scientifically examined and ascertained. Such biological determinism has already always also a normative dimension: it favours human expertise, privileges scientific-technological authority and moves ever closer towards law-like structures of grasping ethics. Whether this position on the elimination of the philosophical category of free will has currency in the long run remains to be seen. What becomes clear, however, is that there is a particular instrumentality prevalent in applied ethics that seeks to establish certain ethical outcomes through regulatory frames, laws and codes. As ethics' modern form in the political context is one of practical concern, chiefly occupied with applying abstracted principles to specific situations, it is useful to engage with some of the underpinnings and critiques of applied ethics to then identify the biopolitical rationales that inform the application of ethics in politics.

'Applying ethics' as a quest for certainty

In recent years, some scholars of philosophy have commented there has been a focus in philosophical thought on the application of ethical principles rather than the ethicality of ethics itself (Raffoul 2008: 271; Bauman 2000: 86). In other words, throughout the past decades contemporary debates have tended to focus on establishing the practicalities and modes of application of ethics, rather than paying close attention to ethics as an autonomous concept that escapes codification or the need for purpose (Bauman 2012: 214; LaFollette 2003: 2). As Rosie Braidotti suggests, this chasm is often encapsulated in the 'thorny relationship' between continental philosophy and Anglo-American approaches to philosophy on ethics. Continental approaches to philosophy, particularly those that fall within a post-structuralist framework, consider difference, alterity, uncertainty and contestability in their discussions of ethics. In contrast to this, analytical Anglo-American approaches to moral philosophy are chiefly concerned with finding principles and solutions within the 'realm of rights, distributive justice or the law' (Braidotti 2011: 300). The latter solicits modes of acting right or wrong with various levels of certainty for the modern subject, while the former 'bears close links with the notion of political agency, freedom and the management of power relations' (2011: 300). While continental approaches resist the schemata of scientific examinations, practical approaches to ethics produce and perpetuate, I argue, scientific-technological rationalisations of ethics. The trend towards ethics as providing a practical guide for moral agents is demonstratively encapsulated in applied ethics as a growing sub-branch of ethics in the hierarchies of political philosophical debates.

Applied ethics[6] has seen a tremendous surge in the past decades and is one of the largest areas of growth in philosophical investigations. As a relatively new area of philosophy, applied ethics remains marked by controversy and debate about what exactly applied ethics is, what basis it should rest on and what methods to approach ethical problems with (Chadwick and Schroeder 2002: 1; Beauchamp 2005: 7–14; LaFollette 2003: 4). The challenge at the heart of the field is exemplified not only in the difficulty of defining applied ethics and its contents,[7] but also by the 'disputes as to what the task of applied ethics should be' (Chadwick and Schroeder 2002: 1). Ethical content in applied ethics remains vague, yet, as Kurt Bayertz observes, 'applied ethics is increasingly being integrated into the training procedures of various professions' and is 'called upon on different levels of practical decision-making', taking on a distinct public role (2002: 36).

Both as a term and as a branch of ethics as a field of inquiry, applied ethics has its origins in the 1960s. It became a relevant approach to ethics in the 1970s with the emergence of bioethics, as a novel sub-field of medical ethics, combining life sciences, biotechnology, politics, law and philosophy. Although the roots of the idea of applied ethics reach as far back as antiquity, the notable increase in the conceptualisation, use of and interest in applied ethics is attributed frequently to wider social concerns over injustices arising in the 1960s and 1970s (Beauchamp 2005: 2). For Tom Beauchamp, it was specifically issues relating to 'civil rights, women's right, animal rights, the consumer movement, the environmental movement and the rights of prisoners and the mentally ill' that gave rise to approaching ethics from a practical perspective, paired with a greater level of interdisciplinary interest in issues of morality (2005: 1–2). Hugh LaFollette similarly attributes the volatile sociopolitical environment of the 1960s and 1970s, where matters of 'racial and sexual discrimination, the war in Vietnam, abortion and the degradation of nature' (2003: 2) became central political issues, to the rise of applied ethics. Furthermore the emergence and pervasiveness of biomedical problems and issues, brought about by rapid technological developments in the medical field and its various realms of applications, were instrumental in boosting the pervasiveness and relevance of applied ethics in a wider social and political context. In other words, applied ethics bears a historical background that already stands in direct relation to the ever-increasing capacity to technologically capture, analyse and understand biological concerns of the human as an individual, and humans as a species.

The turn to applied and practical ethics is thus in itself biopolitically grounded, in that the shift of life into the centre of politics, paired with the technological and scientific capacities for the mathematisation of life, plays a crucial role in the politicisation of *zoe*, as well as the *zoe*fication of politics. It is the calculability of the human self (and the human other), in their physiology, biology, neurology and psychology that allows for a consideration of ethics in terms of physiologically, biologically, neurologically and psychologically established norms of right and wrong. This, in turn, informs the perspective that ethics can, and ought to, provide solutions to calculable problems. LaFollette elaborates in this context the greater accessibility and availability of empirical data as a key reason for the rise of applied ethics, as it is through empirical data – scientific knowledge about the human – and knowledge of the scientific process that ethics can be conceived of primarily in rational terms. He illustrates this relationship with an example:

> We may say, for example, that we should maximize the greatest happiness of the greatest number of people or that we should respect people's rights. However, those claims are little more than vague objects of our homage unless we have some knowledge of human psychology, the nature of human happiness and autonomy and an awareness of the ways that our, others' and institutions' actions shape people's ability to live happy or live autonomously. (2003: 7)

In other words, in order to know for sure what is 'good', we must be able to measure it. While this statement and reasoning are not flawed *per se*, they imply the assumption that there can be a measure of human happiness and that this can be empirically ascertained – in this case through scientific knowledge of human psychology. The utility calculus is of course not something unique to our contemporary condition. Jeremy Bentham famously, and problematically, concerned himself with finding exact measures that would help in calculating quantities of pleasure that would offset quantities of pain with the *Felicific Calculus*. However, this thinking is amplified today as modern information technologies facilitate the collection and systematisation of enormous numbers of experiences captured as data. Such a logics presumes, that the human can reliably be scientifically captured to give content more accurately to practical moral reasoning. In short, it gives credence to the priority of the scientific basis of biological, psychological and neurological humans to establish 'accurate' ethical content, ignoring the very plural, aleatory and uncertain character of the humans in their context, let alone the status quo of science to date being unable to provide any stable account of human nature (Bell 2010: 655). However, it is this calculability that paves the way for ethics to be considered as a possibility for *securing* right and wrong. It is this calculability also that obscures the investigation into the meaning of ethics with a preoccupation of applying a defined set of principles in the encounter with the other in a sociopolitical context of alterity.

Applied ethics, thus, in its various forms, whether bottom up, top down, in coherentism[8] or other methodological approaches, is thus chiefly concerned with finding valid and 'right' ethical solutions that apply to delineated areas of application. The predominance of a rational and very practical approach to ethics has given rise not only to ethical considerations being framed increasingly in terms of dilemma and debate (MacIntyre 1981: 6) but also to a debate that can be won, ethical outcomes that can be secured. Ethics becomes a quest for certainty that the right thing can be, and is, done across the respective field of application; that wrong behaviour is curbed, if not eliminated through the inscription of rules, frameworks and codes that specify and enshrine what the 'right' behaviour is. Bayertz sees the institutionalised role of applied ethics as pointing toward a 'changed purpose for ethical reflection within modern society' and refers to this process as politicisation, whereby applied ethics is understood as 'part of society's problem-solving process' (Bayertz 2002: 42). The tension inherent in this drive towards certainty through the specific application of general principles is evident when we consider the inherently plural and diverse potentiality of scenarios, participants and events for any particular sociopolitical context.

On this basis, Alasdair MacIntyre makes a strong case for the impossibility of applied ethics as a valid regulatory framework for a generalisable 'right' social conduct. His critique addresses a core question that emerges in the context of the practical application again and again: is it possible to apply *general* moral principles, through the use of a regulatory framework, codes or rules, to *particular* social or political situations? In his 1984 article, MacIntyre responds directly to the rise of applied ethics as a branch of ethics, and asks the crucial question: 'does applied ethics rest on a mistake?' He argues that the moral theories or principles that are appropriate to particular situations, in their abundant plurality, cannot actually be *applied*, as each application, if it were true to taking each social particularity into consideration, would have to yield a new moral principle, which leaves nothing to be 'applied' (MacIntyre 1984: 498–511). There is thus an immanent impasse in the relationship of applied ethics and that which it seeks to apply – moral principles – based on the very existence of social particularities. MacIntyre concludes that, as there is no such activity as 'applying' ethics, this new branch fulfils a substitutive role for morality, 'simulacra of moral principles that are what moral principles are transformed into in the great pluralist mishmash of the shared public life of liberal societies' (1984: 511). Moreover, he identifies applied ethics as an ideological measure to lend credence and justification to certain, morally bound, professions, regulated through explicit and implicit codes, as it stands in for an inherently indeterminable morality in a broad social context (1984: 511).

While MacIntyre framed his critique with a discussion of virtue theory, he recognises and addresses two key problems with the then-emerging branch of applied ethics that resonate with a postmodern perspective of ethics: firstly, there is an inherent tension in the demand of applying general moral principles to a potentially infinite range of particular situations in the aim to solve moral problems successfully, and, secondly, there is the nature of the content of morality as something that is essentially indeterminable in the endless heteronomy of potentially arising situations of moral relevance, which must be considered anew time after time in the modern context. Bauman attributes this quest for a certainty of ethical rules and codes to the precarious character of modernity, in which 'the ambivalence of moral judgements' was viewed 'as a morbid state of affairs yearning to be rectified' (Bauman 1993: 21).

Where ethics is seen as guidelines that stipulate what is right and wrong, on the basis of rational measures, the danger is that ethics is either reduced to prudence or seen predominantly in terms of its economy, its costs and benefits (Dupuy 2007: 238–9). The scientific-technological foundations for this are evident. What then can we make of ethics in an international political context as being situated within this branch of ethics that seeks to ascertain ethical conduct through rules and codes rather than contingency and ambiguity?

Taming the infinite: Ethics in politics

Not least since the 1990s, politics has increasingly been considered in its ethical dimensions, domestically and internationally, yet the relationship between politics and ethics – two quintessentially human dimensions – remains somewhat fuzzy and contested (Coicaud and Warner 2001: 5). To gain clarity, it is helpful here to recall the different conceptions of politics, as suggested earlier in the book. Politics proper, in Arendtian terms, is contrary to the politics-as-life-management conceptions of contemporary modernity. In politics proper and its contingent and plural nature, ethics is always already immanent; there is an implicit responsibility we have to one another (Bauman 2000: 84). In contrast, where politics bears a much closer semblance to the professional administration, management and government of populations and their resources, lives and livelihoods, an understanding of ethics as a practical matter for politics is evident. It is in this context that we can understand politics as being grouped, in its ethical dimensions, in the sub-field of applied ethics, in which ethics is understood as embedded in practice, and codified accordingly. It is then a question of 'what kind of ethics should apply' to politics that becomes the core question in such a political realm (Gamble 2010: 74).

While the general field of applied ethics is divided yet again into several sub-categories – medical ethics, bioethics, business ethics and so on – there has to date not yet emerged a category by the name of 'political ethics'. Yet to most scholars in the field there is little doubt that modern political practices fall broadly under the category of applied ethics (Chadwick and Schroeder 2002: 15). Political administration is concerned with real-world questions that demand and seek a practical solution, from issues of housing to matters of criminal punishment, to problems of integration, to issues of birth (i.e. abortions) and death (i.e. euthanasia) and so on, which makes applied ethics as a branch of ethics relevant and useful for politics-as-management. This is similarly true in international politics, where ethics in international relations, global ethics or international ethics, as a relatively new field of analysis, has forged an identity 'as a branch of applied ethics' (Nardin 2008: 594), addressing a wide range of complex international issues from migration to distributive economics, to warfare. To consider international politics, under the sway of applied ethics, as a predominant frame of reference presents, however, the same problems that appear for the broader area of applied ethics, as flagged up earlier. Only in international politics, these problems are amplified by the very plurality of communities and diverse international interests and backgrounds that make up the international political domain.

Not only is the ascertainment of any general moral theory that could be applied to so complex a sphere as the international realm a questionable endeavour but it problematically seeks to use principles external to the realm it deals with in order to solve internal problems. This turns ethics in to a matter of problem-solving, for which a certain level of expertise is required to correctly identify and apply relevant external principles for a correct solution. The framing of moral policies then is reliant on a specific expertise that the politician (de Wijze 2002: 35), the committee (Bayertz 2002: 3) or the analytical philosopher holds to secure the moral content that can be applied. This moral content is then made manifest and sought to be enshrined in laws, rules and regulations guiding national and international ethical decisions. Such ethico-legal constructs are present in nearly all discussions of the ethics of war, humanitarian intervention and, most recently, drone warfare, in which discussions on the legality of procedures and processes obscure a deeper engagement with the ethicality, or the moral content, relevant to the use of violence as an instrument. When the sociopolitical mandate that informs the norm is centred on a calculable humanity and focused on the abstract idea of the survival of life and humankind, it is perhaps not surprising that contemporary theories about ethics, specifically in the context of war and just war theory, are turned into something calculable and predicable, framed as formulas or algorithms with which to determine ethical behaviour (Hutchings 2010: 161).

It is in this context that the limitations of ethics as conflated with cost-benefit utility, or confused with prudence, become problematically evident. Jean-Pierre Dupuy explains the implications of these fallacies and notes that the underlying logic is a scientific-technologically informed understanding of ethics as risk management of sorts. If confused with prudence, ethics becomes little more than a form of identifying and administering ways to avoid costly risks, by which the 'right' thing is done for fear of adverse consequences, not out of duty or responsibility, to put it in stark terms. In other words, the moral deed is considered primarily in its consequential architectural framework, following the economic logic of a 'precautionary principle' (Dupuy 2007: 238). Where ethics is determined as a cost-benefit analysis, the implicit logic has an economic structure – that of risk analysis. Weighing both costs and benefits requires that those can be numerically captured and represented so that the appropriate utility differential can be determined (Dupuy 2007: 239). It assumes that parameters can be known or guessed to mitigate uncertainty. Yet what underwrites ethics for Dupuy is not merely uncertainty but radical indeterminacy. Where the problem of ethics is framed in terms of uncertainty, rather than indeterminacy, it reduces ontology to epistemology (2007: 40). Dupuy's discussion focuses on the relatively young field of (applied) nano-ethics. Nonetheless, his assessment resonates strongly with the wider issues of applied ethics in biopolitical modernity. This is nowhere more evident than in the current debates on the ethics of war, specifically in recent accounts of just war theory.[9] The focus of current just war theory debates lies primarily in finding applicable ethical guidelines, principles and rules to fight justly in wars, and the language and logic employed to arrive at these principles and theories strikingly resembles the econometric approach Dupuy critiques. Where ethics prioritises finding the 'correct' principles for 'correct' solutions to moral problems, ethics is understood as something that can be solved and secured once the right ethical principles have been identified and put in place, enabling moral risk to be minimised. Such an approach reveals a scientific-technological mindset that seems prevalent in contemporary just war theorising. Moral analysis as risk management thus renders ethics a technical matter; a problem that can be solved through processes of abstraction, analysis and optimisation. The language and form of moral analysis is indicative here. Often words like 'cost', 'duty', 'benefits', 'obstacles' and 'expected value', all of which resemble the technical vocabulary of the economist, appear prominently in analytical approaches to just war thinking. And in many cases, this language is paired with the tools of computational syntax, such that moral dilemmas are expressed as abstract problems to which definitive solutions are to be found. This approach to just war thinking reveals a specific idea of ethics as a science, whereby moral philosophy, or ethics, as Jeff McMahan suggests, works 'like a science', in that the working material is 'purified' into neat categories of variables that are independently assessed through algorithmic logics and scientific hypothesising, such that 'authoritative judgments' about conflicts can be reached (McMahan 2013). Such conceptions of ethics emerge from a modern intellectual political fashion of '"principles", "rules" and "theories"', spurred by 'modern science and technology' (Rengger 2002: 360) in which the predominant logic of just war theory (and ethics in politics in general) becomes one of utility and purpose. Ethics, as a technical matter, thus 'mimes scientific analysis; both are based on sound facts and hypothesis testing; both are technical practices' (Haraway 1997: 109). The analytical approach to just war theory lends itself uniquely to formulating ethics in terms of laws. Contemporary approaches to just war theory strongly appeal to law as a structure and source of authority to render abstract ethical considerations as legal principles for the practical application of the laws of war. This focus on law as the supreme ethical reason for practical implementations produces a growing 'legalization of the tradition' (Rengger 2002: 360).

The conception of ethics-as-science is, however, highly contested as it limits ethical thinking considerably. Crucially, in the context of war, the stated aim of this approach is hampered by its own structure (which carries an implicit bias toward technical analysis). It must, by virtue of its scientific-technical format, truncate all those aspects relevant to war and violence that either cannot be easily captured in discrete calculable terms or are deemed irrelevant as a factor to consider. The many numerical values possible for the different levels of harm incurred by civilian populations in war are clearly too multiple to fit into an ethical formula; how do you, for example, 'measure' the loss of a limb, the loss of a family member, trauma and so on? The focus on capturing the number of dead bodies in war (civilian and otherwise) is indicative here. What matters is not the moral relevance of full life experiences but rather the statistically relevant (and manageable) data for biopolitical violence. This carries within it a biopolitical 'flattening-out' of life's rich tapestry of morally relevant experiences to reflect life's *data* only (Dupuy 2007: 248).

Precisely this complicates the moral inquiry into the justness of political acts such as humanitarian intervention, or the lethal use of drones in counter-terrorism practices. In other words, just war discussions are increasingly framed along the lines of practicality and distinct outcomes, and are often confined to the practices of war, rather than the morality and justness of going to war in the first place. For this, analytical expertise and skill in logical reasoning and in conceiving of specific frameworks that can then be enshrined in the laws of war, or the laws of armed conflict, are necessary. It is in this move that the practicalities of war supersede any moral concerns about the wider experiences related to war. Moreover, where cost-benefit calculations are the preferred form of moral reasoning, such reasoning may well be something best done by a computer rather than a human. And indeed, forays into 'machine ethics' are well under way in recent discussions of automated weapons systems and drones – a development I address in detail in the next two chapters.

What emerges from this need for expertise in the conception and the application of frameworks of ethics is the demand for codes that specify how those engaged in matters of life and death (medical or warfare-related) ought to best conduct themselves, so that training can be given and the 'right' moral decision can be ensured. In other words, the emergence of prudence as ethics is firmly instituted here. The conflation of ethical conduct with professional training is prevalent in military ethics,[10] exemplified in William L. Nash's anecdote about a group of soldiers tending to the thirst and fatigue of their prisoners of war. Nash is a retired US Army Major General who served as Colonel in Operation Desert Storm. He describes his arrival at the tactical command post in Operation Desert Storm where he observed a group of forty or fifty Iraqi prisoners huddled together in an enclosure, guarded by two soldiers. It was evident to Nash that the prisoners were hungry, cold and exhausted. Nash describes the scene:

> Before I could act, I saw four American soldiers going toward the prisoners' enclosure with their arms full of blankets and food rations. No orders had been issued, training had taught the soldiers to do the right thing according to the laws of armed conflict. (Nash 2002: 17)

For Nash it is clear that the training in the laws of war has made (and makes) soldiers behave ethically, based on their knowledge of the law and the codes of ethics relevant to their profession. Interestingly, Nash describes the soldiers' behaviour as being admirably compliant with the law, rather than an exemplary act of responsibility and compassion. Here we clearly see the point Dupuy makes about applied ethics' fallacies: professional prudence is assumed in lieu of ethical responsibility.

Professional detachment and prudent conduct in war understood as ethical conduct in war are what engender technologically realised ideas of 'ethical killing', as it is proposed by advocates of lethal autonomous weapons system. Ronald Arkin, roboticist and robo-ethicist at the Georgia Institute of Technology, and adviser to the US military in matters of ethics and robotics, argues that it would make for more humane killing in war if autonomous robotic systems were to be considered as battle units, alongside, or instead of, human soldiers. As alluded to in earlier chapters, these robots would be equipped with an ethics module, programmed with specific information based on the laws of war and armed conflict, so as to eliminate the messiness, ugliness, uncertainty, potentially erratic behaviour and flaws of human soldiers in the battlefield (Arkin 2010: 334–7). His main claim, that robots can be 'better' and more humane in the battlefield, rests on the assumption that robots can be programmed to have emotions as well as heightened rationality, and are able to make faster and more accurate decisions. Robots thus, Arkin argues, have the capacity to make warfare cleaner, fairer, more professional and more efficient. Arkin goes as far as to suggest that, eventually, ethically equipped robots ought to become the guides of mere mortal soldiers in overseeing and adjudicating the rightness or wrongness of their human conduct (2010: 334–7). While this sounds eerily like a storyline for the next sequel of the 'Terminator' series or the plot of a sci-fi novel, the notion that robots are better moral agents and could employ violence in a more humane manner is fast becoming a possibility, although the practical reality of this continues to be fervently contested.

The next chapter examines the foundations and implications of suggestions such as Arkin's in more detail. For the moment it is relevant to stress once more that such a perspective of ethics – one that is based on scientific formulation and professional conduct – reduces the notion of what it means to act ethically to a set of technical guidelines, so as to make ethical behaviour, specifically in the context of war, certain. In short, recent technological developments in warfare and military affairs echo the desire to render ethics finite and concretely definable, codeable and predicable, and benefit from considerations such as Wilson's notion of ethics as rooted in biological functionality. Arkin is not the only scholar addressing the issue of the codeable morality of machines. Academics and practitioners in the field of military technology alike are currently debating not only the very ethics of the use of robotics as substitutes for the human in warfare but also the possibility of creating a formulised 'ethics' that can be implemented into military robotic structures with the lofty goal of being both more discerning and prudent in the act of killing (Abney, Bekey and Lin 2008; Arkin 2010).

Ethical codes in warfare then become a biopolitically informed technical solution for matters of killing, whereby technology, often framed as neutral in character, is the expedient tool for administering the ethico-legal mandate that has been enshrined. The most significant danger lies in the possibility that the application of normalised ethical standards, as given by a scientific authority, eclipses considerations of individual – or indeed collective – responsibility towards others, specifically when ethics is understood as a programme that is to be applied rather than something that arises uniquely, and with multiple demands. I want to be clear here, I am not arguing that the use of codes, specifically codes of conduct or ethical conduct as enshrined through law or other regulatory frameworks, is *per se* a bad thing, or ought not to take place. Rather, I suggest that we should highlight that the resort to codes as an ethical frame of reference transfers responsibility to the regulatory frameworks, and guidelines that have already been established. This, in a plural international context and against a background of technological authority, is problematic, as I will further discuss in the chapters that follow.

The pursuit of certainty for outcomes in politics-as-violence and the truncating of ethics' indeterminable elements for its practical application go hand in hand in this parallel 'professionalisation' of biopolitical practices. It bears repeating here that the political underpinnings are biopolitical underpinnings and not politics in the Arendtian sense. It is politics as administration, politics as people management and population management that facilitates the 'applicability' of ethics, just like business ethics, medical ethics, bioethics and so on. As in other professions, this engenders the need for codes that validate and justify not only the content of the profession but also the general conduct (behaviour – not action) of the participants in the profession. Michael Davis describes the role of codes of ethics for the engineering profession, for example, as follows:

> A code of ethics would then prescribe how professionals are to pursue their common ideal so that each may do the best she can at minimal cost to herself and those she cares about. The code is to protect each professional from certain pressures. (Davis 2002: 250)

Where Davis makes a distinction between the manager, who handles people, and the engineer, who handles technical knowledge, in the profession of contemporary international politics, the mode of handling people becomes increasingly technical, as people are rendered in mathematically and technically ascertainable modes. The biopolitically informed field of international politics, on one hand, gives rise to the use of codes and laws as basis for ethical considerations, while on the other hand also facilitating the ethical justification of practices of political violence under the sway of the *zoe*fication of politics. In other words, when politics sees its *raison d'être* as the regulatory authority of the body politic and its health and well-being, the professional politician, as physician, becomes a useful framework for the justification of violence. The turn to medical language in referring to a body politic or to practices to intervene into it, is illustrative of how condition and image are becoming functionally aligned in the *zoe*fication of politics.

Killing in the name of health

Casting political communities as organisms is not a recent phenomenon in political theory; organic theories of states have been employed in many traditions, going back as far as Plato (Amadi and Wonah 2016: 416). In biopolitical modernity, however, the organic metaphor always bears the potential to see violence as a necessary mode to intervene creatively and professionally, as Arendt warned:

> The organic metaphors with which our entire present discussion of these matters, especially of the riots, is permeated – the notion of a 'sick society,' of which riots are symptoms, as fever is a symptom of a disease – can only promote violence in the end. Thus debate between those who propose violent means to restore 'law and order' and those who propose nonviolent reforms begins to sound ominously like a discussion between two physicians who debate the relative advantages of surgical as opposed to medical treatment of their patient. The sicker the patient is supposed to be the more likely that the surgeon will have the last word. (Arendt 1970: 74)

Arendt's insights resonate today as segments of populations or entire populations are rendered in pathologic terms as 'sick' or 'diseased' and in need of professional help. This is particularly noticeable in the military discourses on the war on terror. Increasingly, medical metaphors serve as a means to assess (diagnose) what is wrong within a body politic, and what can and ought to be done to remedy the ill. This is perhaps no coincidence since the medical profession and the military industry both deal with life and death as their core concerns. Moreover, war and medicine are far more than incidental allies. As Alison Howell points out, since the nineteenth century medical and military industries have shared an imbricated history, which is underwritten by the same scientific-technological logos of managing individuals and population: 'both aim at the population, both are strategic, both claim to produce security' (Howell 2014, 975). Shared terminology such as 'triage', 'targeted killing', 'surgical precision' or 'signatures' is one indicator of the deep technological entwinement of the military and the medical. Similarly, the biopolitical language in which societal assessments are couched is rife with pathologising terms such as sick and healthy, cancer and cure, diagnoses and remedies. The narrative of a 'sick society', a phrase repeatedly used by former UK prime minister

David Cameron after the 2011 riots in London, is illustrative of the moral positioning of the politician as physician.[11] John Keane notes that medical language was instrumental in fascist regimes, particularly Nazi Germany, revealing an obsession 'with unifying the body politic through the controlling, cleansing and healing effects of violence' (2004: 2). Medical or surgical images are, however, not limited to fascist or totalitarian regimes. They continue to function as a veneer to mask the use of violence by and within democracies today by providing the rationale for reforming or removing unwanted elements from society. But while Keane argues that mature democracies see such euphemisms as 'corrupting and contestable', the growing use of such language by mature democracies suggests otherwise (2004: 2). Particularly in military discourses the use of medical terminology has been steadily on the rise since the twentieth century and runs through narratives associated with the Cold War, humanitarian efforts and counter-insurgency programmes (Bell 2012; McFalls 2007; De Leonardis 2008: 33–5). Today, such language is most notable in the war on terrorism, which shapes up quite strikingly along medical terms, whereby 'terrorism' and 'cancer' are addressed in much the same key.

Medical metaphors rely on the anthropomorphic view that communities are organic corporeal entities with human qualities. The state, for example, is conceptualised as a human whose physical strength is epitomised as military power; the welfare and health of this 'person' is typically depicted in terms of wealth and economic prosperity. Such metaphors cast the survival of a community in vital terms, warranting challenges to its military strength and economic prosperity to be seen quite literally as 'death threats' (Lakoff 1991). This analogy rests on a linear and progressive understanding of human development, whereby a community (such as a state or nation) is considered 'mature' when it has been industrialised, and economies that do not function in line with the industrialised model are considered under-developed, or 'immature'. George Lakoff notes: 'there is an implicit logic to the use of these metaphors: since it is in the interest of every person to be as strong and healthy as possible, a rational state seeks to maximise wealth and military might' (1991). Implicit in this is also a moral mandate to uphold the progress of society, understood as health. While phrases such as 'surgical strike' have often been interpreted in rhetorical terms – that is, as a means by which to make a messy war appear cleaner than it is – they also betray the biopolitical underpinnings of contemporary warfare. Metaphors are not neutral in their cognitive effect. More than merely an effective rhetorical device, they are considered by linguists to be a 'figure of thought', which 'consists not merely in *representing* the objects, but in *depicting* them' (De Leonardis 2008: 34). Furthermore, metaphors have the capacity to create a reality, establishing and manifesting a similarity between two concepts where previously there was none. It is through the use of such metaphors, then, that the logos of a certain sociopolitical form is disclosed and circulated.

To see the biopolitical logics at work, consider the language used to describe terrorism today. In Barack Obama's Address to the Nation in December 2015, he refers to the threat of terrorism as a 'cancer that has no immediate cure' (Obama 2015). Such rhetoric is also popular with the current US administration. Donald Trump made use of the trope during his presidential campaign, declaring that the 'cancer of terrorism needs to be stopped before it "festers and festers and only gets worse"' (Belvedere 2015), and the former National Security Adviser to the Trump administration, Michael Flynn, is known to liken all of Islam to a 'malignant cancer' that has 'metastasized' and must be stopped at all cost (Rosenberg and Haberman 2016). Further examples of such rhetoric in current discourse are too numerous to include here, but it is worthwhile highlighting an article in the *Boston Globe* in 2016, which describes in clear terms how strategies of targeted killing (a common term in oncology) for cancerous cells are analogous to combating terror. The article notes that a similar precision strategy for targeting cancer is relevant for combating terrorists: 'Using finegrained analysis of big data, governments could develop algorithms to identify clear markers that distinguish terrorists from the general population', the author writes (Westphal 2016). Here, '[c]ounterinsurgency becomes chemotherapy, killing insurgent cells and sometimes even innocent bodies to save the body politic' (Gregory 2011b: 205). The narrative of military intervention as a 'therapeutic' procedure that remedies the unhealthy aspects of a specific society in turn renders violence as something good and intrinsically moral, and neutralises intrusive technological practices as necessary in the process towards health.

This is strikingly illustrated in a 2010 *Foreign Policy* article in which officers Lt Gen. William B. Caldwell and Capt. Mark Hagerott suggest that Afghanistan could be cured. In the article, titled 'Curing Afghanistan', the authors suggest that Afghanistan is an ailing patient, a 'weakened person under attack by an aggressive infection' (2010). They then explain the logic of US interventions by comparing their role to that of a surgeon, casting the Taliban as an infection for which their counterinsurgency efforts serve as a course of antibiotics. The current advocacy of the use of armed drones in Pakistan, Yemen and Somalia is also couched in these terms, with drones repeatedly referred to as instruments that enable surgical precision in eliminating the cancerous cell that is the terrorist enemy. As such, the use of lethal drones is positioned not only as legal and ethically sound but also wise, a narrative I discuss in detail in Chapter 7.

These examples reflect the contemporary *zoe*fication of politics at work in contemporary military affairs. As De Leonardis aptly explains: 'what is at stake here is a view of society as an organic body that is threatened by some external or internal noxious substances, a *corpus organicus* that can be cured only by a ruler-physician' (2008: 37). In the same vein as two specialist consultants would, Caldwell and Hagerott 'examine' the course of treatment hitherto applied and find that 'Afghanistan's illness' was diagnosed too late. The 'low level antibiotics' employed to date are insufficient to cure Afghanistan. What is needed is a shock to the system; an intervention. It is within this biopolitical logic that stepping up military action is seen as a heavy but necessary dose of antibiotics, which has unfortunate but necessary side effects:

> To be sure, similar to a powerful antibiotic, the use of large numbers of combat troops brings with it side effects that can cause discomfort and pain to the body politic of Afghanistan. The effects range from disruption of civilian day-to-day life to, regrettably, sometimes civilian casualties. Senior NATO commanders seek to minimize civilian casualties and thus apply combat power with restraint and, to the extent possible, surgical precision. (Caldwell and Hagerott 2010)

Emphasised here is the professionalism and expertise of the surgeon who will take lives only when no other pathways are possible, or when accidents happen. An intrinsic moral high ground is firmly assumed by the medical expert in such analogies.

The homology of biological malady and social malady implies a specifically biopolitical power relationship in which the 'governed must submit to the ruler with the same eagerness a patient entrusts his/her health to a physician' (De Leonardis 2008: 39). Caldwell and Hagerott's metaphor is by no means radically new. Conceptions of immunology and virology, for example, were a strong influence on Cold War rhetoric, where 'the germs of communism' were cast as 'a threat to the national body conceived as an individual' (Wald 2007: 172; Schwarz 2016: 68). But, far from abating, medical language is instrumental in US counter-insurgency efforts (see for example Gregory 2008; Bell 2012). Moreover, the discourse of the drone programme takes on a much more radical logic, manifesting Arendt's precocious fears – 'the sicker the patient, the more likely that the surgeon will have the last word' (Arendt 1970: 74). The justification and alleged ethicality of the use of drones is instructive here, as it is consistently framed in terms of the professional expertise of the administrator, the technological expertise of the tools available and the survival mandate that underpins their work. In troubling ways, the rationale behind the defence of targeted killing by drones evokes a modern, militarised version of the Hippocratic Oath. It manifests a perspective of the intervener, armed with lethal drones, bound by professional duty to eliminate a condition of sickness, and now equipped with the 'right' precision tools to do so. The novelty of lethal drone use thus lies in the combination of a medical narrative to justify targeted strikes with the technological capacity to do so.

The underlying rationale of such a techno-biopolitical framing has several consequences for the ethics of political violence. One is that it enables the drawing of dichotomous boundaries between what is normal and what is abnormal, always with a view to prescribing effective treatment for abnormal conditions. This, in turn, manifests a moral prioritisation of necessity and affirms the authority of scientific and technological expertise fit to prescribe a diagnostic and remedial path to that which is designated as in need of a cure. The flipside to this is an ethical demand on the 'diseased elements' in a (global) society to submit to being cured – for their own good, and for that of a wider body politic – by the entity in possession of the knowledge, expertise and technology needed to correctly diagnose and treat their sickness. What emerges is thus a hierarchical power relationship, enabled by the socially constructed medical categories of health and illness, and cemented by a moralised technology (Bauman 2012: 159; De Leonardis 2008: 39). In an ever-greater rendering of science as truth in the inquiry into human biology, the biological knowledge available to the individual becomes limited. The morality of the remedy, understood in terms of utility, stands unquestioned in the survival mandate. The modern aspiration towards the omnipotence of humans means not only that everything is possible but also that we have no choice but to secure the health and well-being of humanity with all means possible. The moment of ethics is thus occluded by the implicit mandate to secure the health and well-being of humanity.

This tendency rests on a reductionist assumption that the complex and contingent elements in a society can be scientifically captured and brought under control – a biopolitical belief that a society can be ascertained and 'cured'. Ben Anderson highlights the way the US military has adopted such methods in their PSYOPS engagement in Afghanistan. These operations hinge on a boundary-drawing process that distinguishes between 'transformable' (curable) and 'non-transformable' (incurable) populations. While members of Al Qaeda, or comparable terror groups, are deemed to be beyond remedy and, from the very outset, must be eliminated, noncombatants and civilians are shifted into a zone of potentiality that renders them subject to ongoing assessments about the level of danger they might pose both now and in the future (Anderson 2011: 224). Just as in preventive medicine, the figure of the 'suspect' becomes subject to preventive actions

– actions designed to secure their healthy and normal behaviour, before it can slip into categories of abnormality. Understanding the population as a collective that has a set of biological characteristics and needs is crucial here. Only when populations are understood as functioning organisms can preventive or predictive technologies work towards a biopolitical goal of elimination. I examine this further in the next chapter.

The medical narrative combines knowledge with authority and thus, paired with technological efficiency, functions effortlessly as a moralising principle in modern societies preoccupied with the rationalisation and application of ethics. The narrative is largely congruent with existing laws and, where it does not fit, the quasi-divine goodness implicit in the profession of the healer and caregiver renders both the intervener and the target of intervention beyond moral questioning. As Lauren McFalls highlights: 'The apparent neutrality of the Hippocratic commitment to human life and well-being, moreover, exempts medical intervention from ethical critique' (McFalls 2007, 1). Morality is integral to the very profession and implicit in all its efforts to save lives. Rendering the target of the intervention in medical terms further depoliticises actions. Either depicted as a victim (the patient) or a symptom of a disease (the enemy), the subject is denied a political dimension on biological grounds. The scientific mandate and the advocacy of technology are crucial aspects in the manifestation of anti-political life politics in which ethical decisions become obscured in the pursuit of certainty.

It is here that Bauman's notion of *adiaphorisation* strikes a chord (Bauman 2000: 92). Drawing on ecclesiastical terminology, adiaphorised acts may be regarded as neither good nor evil, but rather exist in an artificially created amoral space. It is the product of organised modern societies that enables people 'to silence their moral misgivings in order to get certain jobs done' (Jacobsen and Poder 2008: 81). Bauman himself uses the term to denote 'the tendency to trim and cut down the categories of acts amenable to moral judgement, to obscure or deny the ethical relevance of certain categories of action and to refute the ethical prerogatives of certain targets of action' (Bauman 2000: 92). In such a process 'difficult moral and ethical questions are elided in favour of action' (Finn 2014: 1–2). 'Act now!' is the credo, which mobilises both urgency and necessity. Here again, the scientific-technological mandate for administrative processes of modernity comes directly to bear on this process. The technocratic basis of adiaphorisation is represented in the bureaucratic processes that lift issues and relations out of a 'moral space (where the justice and legitimacy of one's dealings with others is always at stake) into the cognitive space of technical performance, instrumental rationality and administrative competence' (Doel 1999: 73). The adiaphorisation of ethical content thus rests on an assemblage of techno-biopolitical and scientific expertise that operates with and on the notion of a body politic as an organic entity. This assemblage of scientific and technological expertise then turns ethical considerations into a technical matter, neutralising these to the point of occlusion. Furthermore still, the ethics of healing an allegedly ill or sick body politic through violent incursions becomes difficult to challenge when faced with this specific techno-biopolitical assemblage. Violence is already no longer the last resort, as the consequences of the violent act are allegedly harmful only to the suspected sick cells. And it is precisely this medical narrative of precision and prevention that allows the US drone programme, for example, to 'penetrate areas and kill people in ways that would not previously have been available without major political and legal obstacle' (Sharkey 2010: 375). The danger is that this adiaphorisation stands to eclipse or at least obscure our moral imagination beyond evaluations of effectiveness or procedural legality.

While the biopolitically informed shift towards applied ethics on one hand facilitates a neutralisation, or adiaphorisation, of ethical content under the sway of the *zoe*fication of politics, it also, on the other, gives rise to the greater reliance and use of abstracted codes and laws for ethical considerations. This condition is particularly relevant in the discussion of violence-as-politics, when the latter is predominantly understood as a branch of applied ethics, and is exacerbated when technological expertise takes on a leading role.

Conclusion

The biopolitically informed branch of applied ethics holds a number of challenges as an adequate framework for understanding the ethics of political violence today. As ethics is rendered a professional and technical subject, it seeks to clarify and ascertain outcomes. This is precarious in international conflict and war. Furthermore, in a *zoe*fying turn of politics, it lifts political acts from the realm of ethical evaluation in framing certain actions as necessary for the health and survival of a body politic. The use of medical narratives is instrumental in this adiaphorising move. Furthermore, it enables the prevalence of codes as ethics, based on a technologised understanding of *zoe*, of humans and their biology. Understanding ethics as a highly rationalised framework informing the rightness or wrongness of the conduct of the human, situated in a realm of predictability, is limited and limiting for a comprehensive consideration of ethics and moral action in modernity and it is important to keep in mind the tension of an ethics that demands a duality of consideration: on one hand ethics is increasingly conceived in terms of its functionality, while on the other it continues to pose questions about the very autonomy inherent in ethical thinking. Applied ethics is an appropriate framework for ethical investigations in matters of politics and violence only when the subject matter itself is considered in terms of utility and functionality. Applied ethics is most fittingly applied in a professional context. For this to be transferable to the realm of politics, where life processes stand at the core, the very elements of the subject matter – politics, violence, the human – must thus be rendered and understood in professional and technological terms.

To more thoroughly consider the ethics of political violence today, it is worth keeping in mind the inherently unpredictable and indeterminable nature of both politics and violence as human endeavours. As Bauman notes: 'Any society is the togetherness of potentially moral beings. But a society may be a greenhouse of morality, or a barren soil' (Bauman 2000: 84). If ethics is shaped by the subjectivities that produce a specific form of society and its political foundations, which then in turn further shape and perpetuate subjectivities, it is useful to look more closely at what type of soil is cultivated by biopolitical subjectivities. In this, the relationship between techno-biopolitical modalities and the human is elemental so that we understand what dominant human subjectivity is produced in contemporary modernity and how this subjectivity enables the justification of violent practices. Specifically, against a background of growing speculative conceptions of a post-human future, the question of ethics is not merely an aesthetico-political question. Rather, it is a question of fundamental ontological importance. The next chapter will investigate the role and place of the techno-biopolitically produced human in more detail.

Notes

1 To illustrate how Arendt might be framed as a thinker of ethics as contingent and incalculable, Elisabeth M. Meade's insightful essay 'The Commodification of Values' (1997) offers a critique of ethics as 'the application of reified concepts like rules, values, and standards' and challenges such an approach as a viable ethical response to situations of crisis (Meade 1997: 124). My own analysis, using an Arendtian biopolitical framework, resonates with Meade's conclusion.

2 The idea of targeted killing as a viable strategy to combat terrorism was first introduced in the US in 1984, by hardliner Lt Colonel Oliver North under the Reagan administration. North drafted the National Security Decision Directive (NSDD) 138, which introduces the idea of pre-emptively addressing the terrorist threat. When the directive was first drafted, North used the term 'neutralise', which was later changed to a less contentious wording, as, at the time, the very idea of assassinations, pre-emptive assassinations no less, was

met with strong moral indignation and considerable legal barriers (Fuller 2014: 778–9; Simon 2001).

3 In this context, Chris Brown quite rightly remarks on the unusually lasting influence of 'essentially medieval theoretical construction[s]' on this particular subject. This has remained unchanged to this day (Brown 1992: 132; Rengger 2013: 7–8).

4 See for example Bergman-Rosamond and Phythian 2011; Bell 2010; Coker 2008; Frost 2009; Hutchings 2010; Kymlicka and Sullivan 2007; Salamon 2015; among others.

5 See for example Willmott 2016; Koch 2012. It is worthwhile highlighting that the scientific tests that contest the notion of free will are heavily disputed; scientific and philosophical scholarship have engaged in a fierce battle over this issue. As with all categories of truth by proof, it would seem that science is on the winning end; the desire to scientifically understand the human brain, and crack the scientific problem of 'free will', continues to inspire countless research projects in neuroscience.

6 The term 'applied ethics' is typically synonymous with 'practical ethics'. For the purpose of clarity I use the term 'applied ethics' throughout.

7 The debate on content centres primarily on the key question as to what the relationship between ethical theory and applied ethics is. Beauchamp goes to great lengths to show that ethical theory and applied ethics should not be considered distinct, while Gert (1984), for example, maintains that there is a separation of ethical theory and applied ethics in a mutually beneficial relationship. Alisdair MacIntyre and others, however, question the possibility of applying an ethical theory to specific contexts altogether. A fruitful and interesting discussion on this topic is captured in a 1984 edition of the *Monist*.

8 As Tom Beauchamp outlines, there are a number of approaches to practical ethics reasoning, none of which is, in his assessment, fully satisfactory. Top-down approaches rely on pre-existing general norms which are applied to new particular situations. The top-down approach is a common approach in applied ethics. The bottom-up approach, in contrast, focuses on 'existing social agreements and practices, insight producing novel cases and comparative case analysis as the starting point from which we commonly make moral decisions'. Coherentism, also referred to as 'reflective equilibrium' or 'coherence theory', favours a reflective approach which starts with an extensive and broad set of moral judgements to build a set of principles to test validity in specific contexts (Beauchamp 2005: 8–10).

9 I am distinguishing, following Rengger, between just war tradition and just war theory, in which the latter denotes the 'scientistic' adaptation of just war traditions in seeking to devise rules and principles from the just war tradition (Rengger 2002: 360).

10 Particularly in the US is a narrow understanding of military ethics as the proper conduct within the parameters of the job prevalent. Other militaries, such as The Royal Netherlands Army, for example, include a more comprehensive ethics training on the basis of Virtue Ethics in their training of military personnel.

11 In August 2011, shortly after the London riots in which mostly underprivi-
   leged parts of the UK capital suffered the consequences of the rioting acts,
   prime minister David Cameron framed the problem precisely in such biopolitical terms when he diagnosed society as being 'not just broken, [but] sick'
   (Cameron 2011a). In his public statements following the riots, Cameron
   laments at various points the decay of moral behaviour and the sickness of
   certain pockets of society. In his value judgement, the behaviour displayed by
   the rioting public was one of immoral (diseased) behaviour, which must be
   met with whichever means necessary, including physical violence (Cameron
   2011b). It thus is only controllable, predetermined 'healthy' behaviour that is
   deemed moral behaviour and all other forms of action that do not meet such
   standards are deemed unhealthy elements that society must be cured of. This
   cure, Cameron emphasises, certainly includes 'first and foremost ... a security
   fightback', greater show of sovereign strength and tougher physical measures
   if need be (2011b). The morality of the cure thus cannot be questioned, as it
   is in making society healthy again that it is employed, thus bearing an intrinsic
   claim to morality.

6

All hail our robot overlords

Und täglich steigt aus Automaten
immer schöneres Gerät.
Wir nur blieben ungeraten,
uns nur schuf man obsolet.[1]
> Günther Anders, 'Molussische Industriehymnen',
> *Die Antiquiertheit des Menschen* (2010: 26)

The previous chapter identified that contemporary conceptions of ethics, like politics and modern ideas of violence, are affected by biopolitical rationales, whereby ethics is rendered as a technical issue for which limiting moral risk and the pursuit of ethical certainty have centrality. I have also shown that, where ethics and professional processes work in tandem, the horizon for the contestation of certain biopolitical narratives is diminished, more so still when technological processes shape moral considerations for the codification of ethics. In this chapter I address the dimension that underpins much of the ethos and mechanisms of biopolitics today – the scientific-technological rationale – and examine how fundamental it is to our contemporary subjectivities. Considering the interdependence of ethics with subjectivity (Campbell and Shapiro 1999: xi; Critchley 2009), it is pivotal that we should take seriously how the scientific-technological condition may produce human subjectivities that shape, if not limit, ethical inquiries in specific ways. It is to this task that I turn my attention here.

In 1941, Herbert Marcuse presciently highlighted what would become an increasingly important question for modern societies: What is the human's relationship with technology as a distinctly sociopolitical force? Writing against the backdrop of a staggering appropriation of technologies for oppression and large-scale violence in Nazi Germany, Marcuse called attention to the fact that, since the technologisation of modern society was set into motion, the world – and the human within this world – has been urged to submit to a rationalisation mandate in which the logic of technology has become a form of social power. As a result, he argues, the twentieth century is marked by a 'technological rationality' in which human autonomy is all but eroded by the instrumental and empirical demands of modern political economies and their inseparable entwinement of the human with technology (Marcuse 1982: 140). Today, this enmeshment of human and technology is visible in virtually all aspects of Western life, and deepening in intensity at an accelerated pace. Technologised societies are governed by this ' "rational apparatus" in which combining utmost expediency with utmost convenience, saving time and energy, removing waste, adapting all means to the end, anticipating consequences, sustaining calculability and security' have become the master mandates of political administration (1982: 142). From new communications technologies to the implantation of microchips into brains to improve performance, to the use of unmanned and autonomous weapons systems, technology and its logic are advancing at a pace that exceeds the political, legal and ethical frameworks upon which we have hitherto built our co-existence in a shared world. While the interplay of humans and machines has a long history, the role of the human in the political economy of the human–technology complex is shifting and the hierarchies are changing. The key characteristic in this assemblage is no longer that of a subject–object relation, or of a mastery of the human over the machine, but rather a co-constitutive merger of human and machine subjectivities, in which neither is fully external to the other. Here again, the analysable, mathematisable human is paramount to the endeavour to understand the human in technological terms, and consequently merge human and machine for better functionality. Humans, perceived as near fully analysable entities in this context, becomes themselves a practical and technical subject matter in this context. However, this harbours a latent danger: wherever human functions can be rendered in mathematical and analysable terms there is a drive to imprint progress in the form of technological enhancements on the patterns identified, rendering humans in their natural flesh, their natural capacity, as flawed and failing.

The co-constitution of biology and technology is a structural signature of modernity, shaping both how we conceive of technology and how we view life. While the human is rendered incrementally analysable, calculable, improvable and replicable through scientific-technological innovations, technology itself is deeply infused with scientific notions of biological processes. The German theoretical physicist Werner Heisenberg clearly identified this modern enmeshment in the late 1950s. He suggested that, in the modern human–technology complex, technology ceases to appear as a manifestation of humans' determination to expand their material capacities, figuring instead as 'a large scale biological process' (1958).

This insight is more relevant today than ever. Writing on the topic of contemporary robotics, Arkin reiterates the relevance of biology for robotics in our time, explaining that '[e]very aspect of robotics is touched by biology ... It's a pervasive influence' (Arkin, cited in Singer 2010b: 90). Where technology and politics blend, the technological subjectivity is thus a biopolitically informed subjectivity, and the biopolitical subjectivity in turn a technologically informed subjectivity. Moreover, it is where both the logic of life (biological processes) and the structure of contemporary ethics (secured as codes) find their logical nexus. Indeed, as Arthur Kroker puts it, 'it may be that human subjectivity has now become so deeply and inextricably embedded with technology that the "question of technology" has now become the question of the human' (Kroker 2014: 17). Where humans are in question, so are their ethics. The socio-political condition we find ourselves in today then warrants closer inspection, in particular in the context of warfare, as technology conditions not only subjectivities but also ethico-political interaction. It is, as Peter Sloterdijk suggests, an anthropo-technical ecology with a particular pervasive twist whereby 'technology puts humanity at risk but will also save humanity by creating superior human beings' (Sloterdijk 2009: 27).

This chapter addresses the question of humans, their subjectivity and their role in the bio-techno-political ecology of contemporary modernity. In this modernity, the human can be enhanced, if not replaced, through technology, specifically in war and warfare, in what Braidotti identifies as a 'new semiosis of killing' (Braidotti 2013: 124), aided by 'tele-thanatological warriors', within a 'technology mediation in contemporary necro-politics' (2013: 126–8). Specifically, I argue that the biopolitical underpinnings of new technologies of warfare institute a hierarchical relationship between humans and the technologies they become fused with, whereby technology becomes a necessary means in the pursuit of improving and protecting human life. Moreover, technology acquires an authoritative dimension as superior rational and functional entity to which humans become inadequate in their functioning. This applies not only to processes of production but also to dimensions of ethics. Where ethics is understood primarily as code, and where technology is understood as an authority that can produce ostensibly superior ethical reasoning, the justification of violence through technology is an evident consequence.

Coding life

I begin this exploration with a fundamental question, pertinent to all politics and ethics: 'What *is* the human?' This may seem like a banal question, but it acquires particular salience as technology merges with biology, physiology, and neurology. It is increasingly less obvious what it means to be human in an age where we are not only living within but also being shaped by a technological universe. The exponentially accelerated pace with which technological incisions into human life advance make a settled theoretical account of 'the human' today impossible. Speed, dynamism and flux are immanent in an ecology determined by technology. As Jacques Ellul noted in 1978, 'Man's environment is completely a function of technology, to which it increasingly adapts itself' (Ellul 1978: 216). This interconnectedness carries deep implications for our understanding of the human body and mind, as well as human interactions with others, which demand perpetually new and ever-faster adjustments.

As Donna Haraway has pointed out, the category of 'life materialises predominantly as information in the 21st century, abstracted through the category [of the] "gene"' (Haraway 1997: 134). Physical and biological aspects of human life are thus defined and rendered in terms of the information they consist of, contain and convey; a process through which biology and computer science become increasingly intertwined. Projects such as the Human Genome Project, which sought to identify all of the 20,000 to 25,000 genes in human DNA, were only part of a broader rendering of life as code. Everywhere, as Ben Rosen of Compaq Computers has put it, 'biology is becoming an information science' (Rosen, quoted in Thacker 2003: 73). This conflation of biology with information gives rise to some fundamental questions about the meaning of human corporeality as information. As Eugene Thacker asks: 'What does it mean to have a body, to be a body, in relation to genome databases?' (Thacker 2003: 73). Moreover, what does this body represent when it is understood under the sway of the Information Technology metaphor? Thinking biology, neurology and information technology in the same register calls into question the very referent 'the human', when defined primarily by their functionality and existence as units of information (Ibid.). Once rendered as informatics, human life assumes the form of so many units and functions, enabling various practices – 'sociopolitical, epistemological and technical' – to be operationalised as means of managing or governing it (Haraway 1997: 134).

Technologies concerned with the production of a better, healthier, more efficient or resilient human, for example, do so by analysing and intervening upon informatic patterns of which life is assumed to consist. Such technologies therefore mirror more basic scientific categories of how different aspects of life might be measured, reproducing the idea that life itself can be grasped in techno-scientific terms. Regarding contemporary understandings of the body and mind as material, Thacker provides a crucial insight here: 'Once the brain can be analysed as a set of informational channels, then it follows that patterns can be replicated in hardware and software systems', opening doors for 'hardware' interventions and virtual replications (2003: 74). Recent advancements in military human enhancements and AI are illustrative here. The work done at Google-owned DeepMind,[2] for example, illustrates this process towards building a virtual human in ever-more detailed replication as the company continues to make advances in building a neural network computer system that mimics human short-term memory in a bid to progress AI towards human levels. At the heart of endeavours like DeepMind's AI forays, the Defence Advanced Research Project (DARPA)'s Peak Solider Performance programme[3] and other, similar projects to enhance and replicate human functions are human processes conceived of as functionalities that need to be improved upon or need to be replaced for better performance. This applies to all observable and measurable aspects of human – their biology, physiology and neurology – so that they can be captured and explained in terms of 'how they work', and 'how they could work better'. Dillon and Reid describe the digital and molecular revolution as paving the way for a new realm of mechanics: 'of life as mechanism and of mechanisms as life' (2009: 57). In the context of biotechnology, this could similarly be formulated as: life as code, code as life. The rendering of life as code hence is essential in order to 'make' life. To imitate, augment and eventually replace human life through technology it must be pared down to its analysable and replicable elements, rendered abstract and graspable – in code. And it is precisely in this mandate of seeking to code human life (life processes, human behaviour) that the foundations of the biopolitically normed human life reside. Life as code, and code as life, carries the immanent biopolitical potential to manifest what is normal and what is abnormal, what is flawed and what is perfection, in all its teleological aspirations, and in all aspects of the analysable human – purely by statistical and mathematical determination. Patterns in information thus become a yardstick for progress in the development of human minds and bodies and their performance. With the human body and mind analysed and rendered as information, the notion of progress toward 'perfection' can take hold.

In this informatic turn, the complexity that characterises physiological, biological and neurological human life is not eradicated through simplification – quite the contrary. Dillon and Reid are helpful in pointing out that comprehending life as information requires also the algorithmic embrace of the 'accident' in the process of evolution and life development, in order to account for and replicate the adaptive and evolutionary powers that human life itself exhibits. Indeed, this embrace forms the very cornerstone of AI and cybernetics. Writing of this 'algebraic life', Thacker puts the possibilities of the body as information as follows:

> [W]hen the body is considered as essentially information, this opens onto the possibility that the body may also be programmed and reprogrammed (and whose predecessor is genetic engineering). Understood as essentially information, and as (re)programmable, the body in informatic essentialism increasingly becomes valued less according to any notion of materiality or substance (as we still see in modern biology) and more according to the value of information itself as the index to all material instantiation – a kind of code for matter. (2003: 86)

With the advent of cybernetics, 'crude and outmoded' biological distinctions of the human could be refined and replaced (Dillon and Reid 2009: 64). Now that life is replicable, not only in its static elements but also in its evolutionary patterns, that which disturbs the process of evolutionary pattern formation in the creation and enhancement of life appears as undesirable or even hazardous. Hence the perpetual latent danger of unruly, unanalysed, and disturbing life in its infinite potentiality and indeterminability. It is this infinite potentiality – the 'pluripotency and totipotency of life-potential', as Dillon and Reid put it – that turns life into an inherent danger for those seeking to control it (2009: 149). As technology's aim is to equalise human capacity (less in terms of political equality, more in the sense of 'sameness'), the inherent adaptive and aleatory quality of human life renders it a perpetual danger unto itself. This, Dillon and Reid conclude, is why securitisation must necessarily be future-oriented – it is a fundamentally prophylactic outlook, carrying an immanent prevention mandate.

The goal of the technologisation and informationalisation process is a form of homeostasis, a kind of ordering and stabilising principle for information patterns. The result is a more purified life than unaugmented life can ever be, in the pursuit of perpetual perfection, by means of scientific-technological embodiment, whereby the anthropomorphisation of technology, specifically in its alleged autonomy, becomes explicitly manifest. The produced, improvable life is, by default, the better life. This is at once both a philosophical evaluation – about what is better and what is worse life – and a functional imperative to measure and improve upon humans as computational units. The *value* of 'traditional' humans in their embodied corporeality resides in the informational processes they offer, so that technology can code and thus improve these, mimicking and moving beyond them in the evolutionary process. This does not render humans obsolete so much as reconfigure their relation with technology, which now appears as a form of authority by virtue of its functional superiority

(Halberstam and Livingston 1995). The rational, computational mandate of technology is emphasised in such a perception of human life, wherein computational essence has priority over other, less ostensibly rational, aspects of the human. It is easy to see then how non-calculable aspects – emotion, intuition, compassion, and so on – need to be truncated to serve the evolution towards perfectly ascertainable life. As we have seen with rational ethics, when life is captured in measurable terms only, the full experience comprised in life is necessarily truncated. In other words, where life is conceived as code, that which cannot be coded falls necessarily outside the realm of the manageable and thus politically relevant.

This raises another question: What is the imperfect human in relation to the potential perfection of technology? And what does this do to our idea of moral responsibility?

Techno-fatalism

The social and cultural relationship between 'man and machine', technology and human life, has been the subject of a cornucopia of sci-fi culture, from Fritz Lang's *Metropolis* to Isaac Asimov's Robot Series to *Robocop* and not least the 'Terminator' franchise. Since 2010, the subject of cinematic exploration gets darker and more complex, with synthetic humanoid life forms outwitting and outgrowing mere mortals at every turn.[4] Encapsulated within these and other works within this genre is a certain anguish over the possibility of machines 'going rogue', becoming independent, intelligent mechanisms that acquire human life capabilities in ever-more realistic shape but with superhuman strength, endurance and computational capacities, eventually spinning out of control on a violent path of destroying all that is human at worst, or simply having no more need for humans at best (Singer 2009; O'Connell 2017: 105–7; Bostrom 2014). The underlying question addressed and played out in these fictional accounts of a robotic dystopia is whether humans can and will (and now also *should*) remain masters over the machines they have created.

With technologies that comprise progressively greater complexities, making machines ever-more lifelike and increasingly autonomous and intelligent, capable of making decisions, the question of the hierarchy of humans and their machines remains unanswered and underexplored. In a fervent drive for progress, scientists and roboticists work feverishly to replace what we hitherto have known and understood as human life with bigger, better, bolder artificial versions of what life ought to be – fully acknowledging, if not embracing, the possibility of becoming outmoded as humans. Today's machines are designed to outpace human capabilities. In contrast, old-fashioned human organisms lack comparable processing capabilities and might, eventually, 'face extinction' (Singer 2009: 415). Echoing this anxiety, technology tycoon Elon Musk has issued a dire warning about the dangers of rapidly advancing AI and the prospects of killer robots as capable of 'deleting humans like spam' (Musk 2014; Gibbs 2017). Musk is not alone in his cautious assessment. In his seminal work on military robotics, *Wired for War*, Peter Singer extensively documents this attitude among the robotics and military experts he has interviewed in the course of his research. There is a palpable sense among some scientists of the risk that technology might be spinning out of control at some stage, outpacing human intelligence exponentially. In this context, Singer quotes the military robotics expert Robert Finkelstein, who states in no uncertain terms that it is in the nature of robotics that machines will eventually outperform humans, and Finkelstein all but surrenders to his own uncertain future: 'It could end up causing the end of humanity or it could end war forever' (Finkelstein quoted in Singer 2009: 415). Other technology experts echo this sentiment. The nano-technologist Eric Drexler speculates that, unless we understand the technology we create, 'our future will likely be both exciting and short' (Drexler quoted in Singer 2009: 415). The philosopher Nick Bostrom reiterated this sentiment in a 2015 UN briefing, and highlights that AI may well pose the greatest existential risk to humanity today (Bostrom 2015). Stephen Hawking, Bill Gates and others have joined in the cautionary chorus that current developments in AI, paired with robotic mechanics that could potentially be weaponised, may, in the not-too-distant future indeed pose the greatest existential risk to humankind (Sainato 2015).

Statements like these betray a certain fatalism on the part of humans who have, in fact, invented, designed and realised said autonomous machines; they also pose the question whether the advancement of technology can indeed still be considered a human activity or whether it has moved into a post-human sphere that we are yet to comprehend, or may indeed never be able to comprehend. Humans, as conceived by contemporary post-human discourses, are in the conscious process of perpetually overcoming themselves though technology (see, for example, Braidotti 2011; Cudworth and Hobden 2011; Agar 2010; Papadopoulos 2010; Diprose 2009 among others).[5] The term post-humanism is far from being fixed in literature and scholarship. At the broadest level, the term denotes a shift away from an anthropocentric way of thinking to include plant life, animal life and materialities in sociopolitical considerations. There are many forms of thinking 'post' humanity; the technological enhancements of humans are one aspect – often also referred to as trans-humanism. For trans-humanists, the human is 'a work-in-progress', perpetually striving towards perfection in a scientific-technologically facilitated evolutionary process to leave behind the 'half-baked beginning' of contemporary humanity (Bostrom 2005: 4). Trans-humanism, however, is, as David Roden points out, underwritten by the technologically facilitated drive toward better *human* life. It is, he notes, a normative position whereby trans-humanists advocate the freedom to self-design through technology (Roden 2015: 13–14). Trans-humanism, Roden argues, 'is thus an ethical claim to the effect that technological enhancement of human capacities is a desirable aim' (2015: 9). However, the pursuit of trans-humanism through AI, nanotechnology, biotechnology, information technology and cognitive sciences (NBIC) and other scientific-technological interventions does not appear to offer a stasis. Rather, in current developments, the trans-humanist 'meta(l)morphosis' (Braidotti 2011: 77) of human and machine appears to move towards 'an explosion of artificial intelligence that would leave humans cognitively redundant' (Roden 2015: 21). What had been a normative position of trans-humanism then becomes, in Roden's terms, a speculative post-humanist futurism, which merely makes a metaphysical claim about what the world *could* contain. In such a speculative post-human future, there may well be no human at all.

This perhaps unintended move towards a purely speculative technological future harbours an implicit paradox in two aspects. The conception of science and technology as outmoding humans and their capacities is inherently a human activity – it is not determined and initiated by an implicitly non-human entity which demands or elicits submission on the basis of their philosophical autonomy, rather it is through human thought and imagination that this context emerges in the first place. The human is thus always already immanent in the technological post-human. Furthermore, within the post-humanist credo still lies an implicit humanism. Thacker identifies this appropriately when he explains that the post-humanist approach is one where 'technological progress will necessarily mean a progress in "the human" as a species and as a society' (Thacker 2003: 75). This is useful to keep in mind when considering the implications this has for the moral authority narrative of technologised warfare addressed below. Concomitant, as the Janus face of either, a trans-human or a speculative post-human future, is a growing understanding of 'traditional' humans in their unaugmented capacity as flawed, error-driven, slow and fallible. In other words, the relationship of the human and technology harbours a peculiar duality of aspects. One is marked by scarcely mitigated techno-enthusiasm pushing toward overcoming human life, so as to realise a greater richness in cognitive capacities that currently lie beyond the limitations of a 'traditional' human body and mind.[6] Here too, the normative dimension is evident as more fervent advocates of this trans-humanist position, like the futurist Ray Kurzweil and roboticist

Hans Moravec, consider the acceleration to an ever-faster overcoming of human limitations through artificial materiality and intelligence not just a normative position but a 'moral imperative' (Coker 2015: 117). Implied here is a techno-optimism, informed by what might be considered an anthropo-pessimism. Such a techno-optimistic perspective resonates with Arendt's astute critique of (biopolitical) humans as perceiving themselves as the makers of all things, specifically life and history in modernity. Such a perspective comprises the desire and the drive not only to improve our own functionality as humans into flawless, better, stronger, more perfect and more divine creatures ourselves but also to *make* life better, more resilient and functional. The creation of artificial life, not only as a creation of life forms as such but as an improvement on existing life forms, presents the culmination of humans as perceiving themselves not only as in the image of God, striving toward infallibility and perfection, but like God, in a creative capacity. It is also in this aspect that the medical narrative of the human as (medical) expert in the decision to operate on a sick body politic resonates; the quasi-divine human with the knowledge, skill and technology to decide authoritatively over life and death. What crystallises is a peculiar relationship of the creator and their offspring – artificial life – who will one day supersede those that have given life to them.

And herein resides the other side of the coin in the emergent duality in the relationship between humans and machines: that of total surrender to the quasi-human robotic technologies that have been created, in a fatalistic passivity, bordering on the suicidal, accepting what is apparently inevitably to come. As Moravec notes with an almost audible sigh: 'I've decided that it's inevitable and that it's no different from your children deciding that they don't need you. So I think we should gracefully bow out' (Moravec quoted in Singer 2009: 415–16). The hierarchical relationship in which artificial machine systems and computational logics supersede humans in authority is evident in many facets of contemporary modern life, including our relationship with GPS navigation systems, our growing reliance on digital technology to identify if our physiological metrics are normal, our greater trust in the veracity of big data or our progressive dependency on a range of technologies to make better decisions than we could make ourselves alone. Consider, for example, self-driving cars and Gary Marcus's suggestion, published in *The New Yorker* in 2012, that autonomous machines, particularly driverless cars, might become inherently more moral, as they function better than humans when it comes to driving (Marcus 2012). In turn, the decision to drive a car, rather than let the car drive itself, might be seen as immoral. Here, technology relates paternalistically to the human, because we, as humans, are confronted with our functional limits. This in turn produces a dependency on the techno-paternalistic relation. Such forms of techno-paternalism also inform the push towards the development of lethal autonomous weapons systems, and the problematic notion that they are able to make more ethical decisions in the battlefield.

Techno-paternalism

This new hierarchical positioning of the human and technology represents a shift in the political economy in the relationship. The 'creator' of the machine accepts the position of being inferior in relation to technological forms of authority (be these robots, cyborgs, bionic limbs, health apps or GPS systems). This surrender relies on the assumed techno-authority of produced 'life' on one hand, and an acceptance of inferiority – as an excess of the human's desire to become machine – on the other. The inherently fallible and flawed human will never be able to fully meet the standards of flawless functionality and potential perfection that is the mandate for the systems and machines that are created artificially.

But it is here, in this hybridity of being simultaneously deity (creator) and mortal (human), parent and child, that an inherent tension resides. Günther Anders diagnoses the condition that ensues in the switch from *creator* to *creatum* as 'Promethean shame' (Anders 2010: 25–95). For Anders, the relationship between modern humans and their machines is characterised not only by a role reversal in which the human creator of the machine becomes subject to the machine's flawless powers. It also implies shamefulness about not-being-machine, encapsulating awe and admiration of the superior qualities of machine existence and the promise of flawless perfection for a specific role or task (2010: 27–30). This is a distinctly modern condition and reflects the complex entwinement of human labour and machine labour from the industrial revolution onwards. In this condition, modern humans measure their worth and general moral standards by the flawless functionalities of machines, yet forever realise that, despite being producers, they themselves cannot fully be products, never fully live up to the strength, speed, precision and functionalities of their creations, and thus never fully fit into the normed environment that is shaped and determined by ever-accelerating technologies (2010: 30–75). Anders, writing on the notion of Promethean shame in 1956, sees precisely in this emerging shame of the human in an increasingly technologised modern context, the advent of the drive for human engineering, and a desire for physio-technology that aims to first determine, and then to overcome, the restrictions and limitations of the human body. The ultimate goal is to not merely to interpret humans and their function but to change them, to render them *made*. Anders sums up the desire to become technology in relation to a sense of flawed humanness:

> The desire of modern man to become self-made, to become a product has to be considered faced with this changed foil: not because man does not tolerate anything that is not made by him does he want to make himself, but rather because he does not want to remain un-made. Not because he is ashamed to be made by others (God, gods, Nature), but because he is not made at all and as un-made man remains inferior to his own fabrications. (Anders 2010: 25)

The notion of shame is crucial: I consider shame here in line with what Helen Lewis calls 'overt shame', akin to a 'feeling experienced by a child when it is in some way humiliated by another person' (Giddens 2003: 65); shame as an instantiation of being exposed, shame as 'a concern about the body in relation to the mechanisms of self-identity' (Giddens 2003: 67). Corporeal shame and identity shame are bound up with the human–technology complex in modernity. To compensate, adapt to and fit into the technologised environment, humans seek to become machines through technological enhancement, not merely to be better but to meet the quasi-moral mandate of becoming a rational and progressive product: ever-better, ever-faster, ever-smarter, superseding the human limited corporeality, and eventually the human self. This not only adheres to a capitalist logic, shaping subjectivities accordingly in line with values of production, but is, in its subjectivity-shaping capacity, deeply technologically informed.

These developments harbour two important consequences: one relates to the ontological security of the individual in a technologically dominated society, the other to the potential for social and political control – the two are related. The first aspect is that human-focused technologies present a rather unsettling potential for a shift of human capacity into a regressional state of incompetence and impotence and alters the understanding of one's being in a technology-driven world. This gives rise to latently lingering questions of ontological nature: can I even still trust my own biology or psychology? Can I trust my own body – can I even know my own body or mind if functionalities are technologically enhanced and now fall outside of the remits of what I can comprehend?

This translates into a context where humans, in their precarious condition, have but little choice to seek a source for authority, in a political and moral sense, in science and technology. And it is precisely here that the potential for biopolitical control in the wider sense can gain its foothold. Technological developments of scientific minutiae on biological human life create a higher dependency on said technologies in aspects of day-to-day life. Once a reliance on the constant and uninterrupted availability of these enhancing and augmenting technologies for day-to-day routines has been cemented, the disturbance of such routines can put the ontological security of the individual (and populations) in jeopardy. And, following Giddens in his assessment of self-identity in modernity, 'in situations where the ontological security of individuals is put in jeopardy by the disruption of routines, or by a generalised source of anxiety, regressive forms of object-identification tend to occur' (Giddens 1987: 218). In other words, with an ever-greater drive toward a synthesis of human biology, psychology and physiology with enhancing and superior technology, the issue of what role humans have in such a context becomes increasingly precarious in an ontological and political sense. As David King notes, 'when organisms to be "improved" are human beings, these issues become not just ethics and political, but existential' (King n.d.).

This move towards a technologised understanding of the self hence has socio-political consequences. With Marcuse, we can understand this development as a radical rationalisation of society: 'The world had been rationalized to such an extent and this rationality had become such a social power that the individual could do no better than to adjust himself without reservation' (Marcuse 1982: 142). Applying biopolitical (although not explicitly so) frameworks of prophylaxis and virulence, Baudrillard draws a connection between the increased cerebrality of machines and a subsequent auto-immunisation ability of bodies. Dispossessed of the capacity to count on their own antibodies, he attests, the human becomes increasingly 'vulnerable to science and technology' (Baudrillard 2009: 67). In seeing his metaphorical assessment through, Baudrillard too arrives at a condition of techno-paternalism when he likens the immanently vulnerable human life to the condition of the 'boy in the bubble', who is '"cuddled" by his mother through the glass, laughing and growing up ... under the vigilant eye of science' (2009: 68). The parenting, as Baudrillard concludes, is here done by science and technology – computers. Baudrillard's inference of this condition in relation to society is lucid. The greater our augmentation and alteration of life with technological prostheses and substitutes, the greater our necessity to submit to the superiority of the artificial proxy, which immanently holds a technologically informed ordering principle upon which society orients itself. This relationship of creator to creation and the concomitant (and apparently inevitable) role-reversal of parent and child inevitably also inform social and political structures in a very real biopolitical sense. Baudrillard detects a proportional relationship of this act of technological substitution, not just in light of individual life but in society as well: 'The social system, just like the biological body, loses its natural defences in precise proportion to the growing sophistication of its prosthesis' (2009: 70).

By accepting the role and place of the human as inferior to the anthropomorphised authority of technology, we thus not only submit to a hierarchical relationship but are also complicit in the normalisation and perpetuation of the narrative of the always flawed and imperfect human in need of technological support from those with the expertise to provide it. In this, priority is given to the creating authority and its creation in a continual and ever-accelerating cycle. Furthermore, once what is normal has been reframed in technological terms, based on the informatic essence of humans, the potential to render some particularly uncontrollable and unanalysable elements of (global) society as lethal and in need of not merely identification but eradication, facilitated through new death machines, is made possible.

Abstracting war

The techno-authority that underwrites contemporary modernity is a feature of current modes of warfare too, in that technologies inform and feed into the streamlined logos of contemporary war as computable. This scientific-technological lens has been advanced for US military engagements since the Vietnam war, in which 'formal and mechanical modelling provided the categories and techniques by which the Americans understood themselves and the enemy' (Coker 2008: 38). The scientific-technological perspective draws on different fields of knowledge and practice, ranging from mathematical models, systems analysis, gaming, to cybernetics, using these to inform war strategy at elite levels. Such an approach to war, and its ethics, truncates 'experiential and situated knowledge', along with everything else that cannot be rendered in quantifiable or calculative terms (Coker 2008: 40). The upshot is a prioritisation and privileging of all quantifiable aspects in and of war. This scientific-technological perspective also underwrites much of the revisionist just war thinking, as discussed in Chapter 5. An obsession with discrete numbers in deliberating the permissibility of harm to civilians (number of deaths are often the focus here), and the fixation with data, reflect this priority. Technologies for identifying potential collateral damage estimates are a crucial tool in assessing the incidental killing of civilians, helping to ensure that campaigns may stay within legally accepted bounds. However, as discussed in the previous chapter, here too the methodological rationale is often a matter of form over substance. Technological authority is pivotal in this process. In her analysis of necessity in just war reasoning, Neta Crawford highlights precisely this point:

> The technical analysis is used to help decision makers stay within the law, but it may also serve to excuse decisions that we might otherwise believe were wrong and to defuse the moral responsibility for actions. The moral tension between military necessity, and discrimination and proportionality are not eliminated, but they are smoothed by the use of technical analysis. In a sense, some amount of authority over jus in bello was ceded to the military, then to military lawyers, is then ceded to technical analysis and a form of computer-assisted expertise. (Crawford 2013: 233)

In other words, the technological capacity factors in as a superior calculative information provider, which is ostensibly neutral in character and, at any rate, superior to the human. A techno-authority is thus implicit in most contemporary wars conducted by US and allied militaries. It is worthwhile to stress here that this authoritative relationship has an effect on our agency and ability to contest technological decision-making. In his discussion of surveillance and predictive policing systems, Kevin Miller highlights the uncritical trust placed in computer-based systems. He notes:

> In decision-systems, study after study across numerous disciplines has confirmed the phenomenon of 'automation bias [that] occurs in decision-making, because humans have a tendency to disregard or not search for contradictory information in light of a computer-generated solution that is accepted as correct'. (Miller 2014: 122)

This applies not only to purely automated decision-systems but also to 'mixed-mode' systems where a human is 'in the loop' to review the decisions. His point is that, even though there are humans in the process, there is little chance of errors being reduced or challenged once a decision has been made, because the computer wields a tremendous authority as superior decision-maker. This effectively makes the contestation of an unethical algorithmic calculation, for example, highly problematic. Here again, expertise and technology meet in a powerfully commanding merger. As technology and biology become further entwined, and the logos of the human is predominantly framed as techno-logos, the horizon for contestation of technological decision-making is diminished and techno-paternalism finds its foothold. This problematic is substantiated in both the drive towards military human enhancement for better and more resilient soldiers, and the move toward developing and using lethal AWS and military robotic AI, to accompany, and probably eventually replace, soldiers in action. Both developments posit the human as a weak, insufficient element in the endeavour of war, and both institute a normalisation of a techno-regime for 'better' warfare, and as such cast ethics in war in a technologised light. In this final section, I will only briefly touch on military human enhancement to lend emphasis to the increasingly intrusive scientific-technological conditioning of fighting bodies in warfare and the technological conditioning this produces.[7] My main concern here and in the next chapter is the production of technologically informed justifications of violence and the emergence of new thanato-technological regimes of warfare.

Engineering hearts and minds

Military spending on robotics is at a record high globally, and is projected to grow rapidly, from US$5.1 billion in 2010 to US$16.8 billion by 2025 (Sanders and Wolfgang 2014). While the number of countries producing and owning UAVs has skyrocketed since 2010 – an estimated ninety states are currently using them in some capacity – the development and proliferation of AWS is fast becoming the new frontier, with many commentators fearing a global arms race for autonomous and intelligent robotic systems. The global interest and investment in neuro-, bio-, information and cognitive (NBIC) technologies for the enhancements of soldier bodies and interference with enemy bodies is similarly thriving. DARPA currently runs several projects, which aim to enhance, augment or supplement soldiers' physical, cognitive, sensory and metabolic capacities to render individuals stronger, more durable, more resistant and attentive in warfare (Galliott and Lotz 2015: 2). From neurological, cognitive and biomedical augmentation to body armour and smart exoskeletons to fortify soldiers on the ground, high-tech militaries work to make their soldiers and personnel increasingly entwined with technologies in progressively intrusive ways for greater resilience, better performance and, importantly, fewer deaths. This technological fortification serves several goals. At a most basic level, technological enhancements aim at reducing the vulnerability of a soldier's fragile and limited physical constitution. In a military organisation where every soldier's death is a near-prohibitive political cost and directly contradicts the survival mandate, vulnerability presents a core problem. Making soldiers 'kill-proof' is thus the aim of DARPA's initiatives (Singer, quoted in Abney, Mehlman and Lin 2013: 6).

The assertion of the human as a flawed, fallible and ultimately inadequate factor for successful contemporary warfare dominates the development of new military technologies. The conviction that drives programmes such as DARPA's Peak Soldier Performance (PSP) initiative, or the growing use of automated and unmanned technologies is the diagnosis that '[a]t present, the warfighter does not processes the physiological and psychological capabilities to keep up with the advances in technology' (Holloway and Gruber 2003: 1). To mitigate this, short-term enhancements such as the TALOS (Tactical Assault Light Operator Suit) are designed to boost the physical capabilities of the wearer (speed, strength, agility), while protecting the vulnerable flesh (Coker 2015: 55). The promotional video for Revision Military's 'Kinetic Operations Suit' is instructive, as it claims: 'rely on the human body alone and you may need to pick between mission, and safety. Combine innovation and the human body, and you have an unstoppable capability' (Gallagher 2016). Similarly, medium- and long-term enhancement measures as intended in the PSP programme include invasive biotechnological procedures and pharmacological augmentations so as to amplify soldiers' energy levels, alertness and metabolic capacities to enable them to operate at peak performance levels for an extended period of time (Abney, Mehlman and Lin 2013: 7). In other words, programmes such as these aim at tuning the human body up to function like technology, or, at a minimum, to adhere to the logic of technological performance. The unenhanced human soldier body is consistently framed as the weakest link. In order for the body to function in adherence with machine logic and technological power, human limitations must be overcome, whether that is limitations to visual capacities, sensory abilities, endurance or simply the soft fleshy condition of the mortal body. In this way, the soldier is progressively conditioned towards becoming an operative system, as part of a technological weapons complex. The technological hardware system, however, is the benchmark and authority for optimal performance.

It is evident that none of these new technological endeavours and aspirations would be possible if the human was not conceived primarily in terms of biological and neurological processes that can be ascertained and analysed for an effective amplification and intervention. Such epistemological foundations also serve to direct new technologies toward an enemy or enemy populations. Initiatives such as DARPA's 'Battlefield Illusion' focus research efforts toward neuroscientific and bioscientific technologies to better understand how 'humans use their brains to process sensory inputs', with the goal to '"manage the adversary's sensory perception" in order to "confuse, delay, inhibit or misdirect [his] actions"' (Shachtman 2012). Advances in research of receptor systems in humans could, for example, lead to the synthetic production of airborne bio-regulators 'that can cross the blood-brain barrier and induce a state of sleep, confusion, or placidity, with potential applications in law enforcement, counter-terrorism and urban warfare' (Tucker, quoted in Royal Society 2012: 49). The human in warfare becomes the object of analysis for the potential of biological and neurological interference and the subject of targeting, based on this scientific knowledge. Here again, we see a capturing of the human as primarily a functional entity of biological and neurological processes at work, by which desired outcomes can be engineered. This limits other, political, approaches. A 2008 report on *Emerging Cognitive Neuroscience and Related Technology* poses the questions relevant for the US DoD thus: 'How can I know what people know?', 'How can we make people trust us more?' and 'Is there a way to make the enemy obey our commands?', among others (National Academies 2008: 16–17). Technologies such as these do more than help war makers narrate their efforts as clean, increasingly risk-free and above all scientific endeavours. They also serve to radically alter the role and place of the human in modern warfare. A political approach that would focus on 'winning hearts and minds' is no longer a priority when the suasions of hearts and minds could possibly be engineered by scientific-technological interventions. Most of the enemy-directed NBIC interventions rest on ambiguous legal grounds. Morally, they open new and important horizons about violence in war. My analysis here is focused primarily on US programmes; however, many other states are developing similar technologies at present. Notably China, Russia, India and Israel are all pursuing military NBIC technologies, and will probably contribute to shaping the contours of global warfare to come (Galliott and Lotz 2015: 2).

While the invasive and augmentative technologies highlighted here move clearly toward positioning the human as an object which is to be improved upon, new technological ideas and advancements in autonomous and intelligent weapons systems challenge ideas of agency, limit contestability and potentially produce new regimes of killing in war.

Robo-wars

Among the most contested new technologies in military operations today are automated, autonomous and increasingly intelligent weapons systems. Drone technology and its capacity to administer lethal force from great distances are often seen as having provided the gateway for ever-more distancing and autonomously functioning weapons systems to be designed for advantages in warfare. And while the debates about whether autonomous systems should be designed and used to make kill decisions are heating up, the development of ever more sophisticated and intelligent AWS appears to continue unfettered. Putting to one side the popular sci-fi narrative of 'robots-gone-rogue', it is evident that for the wider public, and a growing number of robotics experts and scholars, there is something deeply unsettling about the idea of autonomous robotic weapons system doing the violent bidding of soldiers in war.[8] An international study conducted by the Open Roboethics initiative has established that a majority of survey respondents (67 per cent) are in favour of an international ban on all lethal AWS, and the number of campaigns and policy groups addressing the dilemmas associated with their development and use is rapidly growing (Open Roboethics Initiative, 2015).[9] At the same time, however, there is a notable drive to push the project of AWS forward among military decision-makers, roboticists and other technophilic commentators.[10]

To date little is known, however, about how far advanced these systems are, nor about who is in possession of them. The embroilment of commercial and military technologies in this development makes it difficult to obtain a clear picture on this. What we do know, however, is that the United States, the United Kingdom, South Korea and Israel have systems with varying degrees of autonomy in their respective war chests (Koch and Schöring 2015; Carpenter 2013). Proponents and advocates of AWS technologies can be found across the board, within engineering and roboticist circles as well as amongst military officials and university professors. US military advocates stress that such systems yield considerable benefits in terms of greater service personnel protection, lower staff costs, more lethality and higher levels of predictability (Zacharias 2015). Further motivations include their alleged potential to 'extend the warfighter's reach' and reduce the number of friendly fire casualties (Arkin 2015). Crucially, Arkin argues that AWS can produce more ethical behaviour in war, by reducing war crimes or human atrocities, and by providing moral guidance for soldiers. This argument has caught on among a wider group of experts and scholars. Evan Ackerman, for example, suggests that, since the development and use of AWS is already under way, we might as well focus on their numerous benefits, such as their imperviousness to emotion or fatigue, or their ability to follow Rules of Engagements (RoE) more stringently than a human solider could. For Ackerman, these benefits constitute an ethical mandate for militaries to choose such systems over humans where possible (Ackerman 2015). Opposing these claims is another set of positions that stress the ethical challenges and moral dangers posed by lethal AWS in particular. The 2012 Human Rights Watch report *Losing Humanity* is paradigmatic in this regard. The report makes a strong case against lethal AWS on the grounds that they are 'incapable of meeting international humanitarian law standards', as they lack the human judgement necessary for a decision in compliance with International Humanitarian Law (Human Rights Watch, 2012: 3). The report further highlights the problem of accountability (who is responsible for a technological malfunction resulting in loss of innocent lives?), as well as the dangerous lure of risk-less and inexpensive war.

Unlike drones, AWS are systems in which the human operator is no longer 'in the loop', or even 'on the loop', but possibly entirely out of the loop. AWS are autonomous systems with the potential capacity to 'identify, track and attack humans or living targets' (Sauer 2014). The most commonly used working definition of AWS is the DoD 2012 directive 3000.09, which considers AWS as: 'A weapons system that, once activated, can select and engage targets without further intervention by a human operator'. This is a relatively broad definition, which leaves a wide margin for both the level and the meaning of 'autonomy' for such systems. At this time, much of the future capabilities and use of AWS are still a matter of speculation and crystal ball gazing. It is nevertheless clear that the desire for and drive towards implementing an ever greater level of technological autonomy in military operations is well under way and lined up to fundamentally alter the character, modalities and perspectives of warfare. Autonomy is thus a 'game-changer', set to rewrite the role, status and involvement of human life on either side of warfare. Consider, for example, the DoD Autonomy Roadmap, presented in March 2015, which sets out the agenda for greater levels of machine intelligence and learning (MPRI), as well as rational goals for human–machine interactions and collaboration (HASIC), and a concept of 'Calibrated Trust' intended to give the human an understanding of 'what the [machine] agent is doing and why' (Bornstein 2015: 14). With the drive for greater technology autonomy comes the apparent desire for greater technology authority. 'Human-autonomy teaming' is a partnership in which the human, at least ostensibly, still decides when and how to invoke technology's autonomy (Endsley 2015). Whether this is possible in all contexts, such as those where humans lack the sensory capabilities required for the task they are undertaking, is very much a question that needs asking. On a not-too-distant horizon is the spectre of intelligent machine autonomy, equipped not only with superior sensors and agility, but also with reasoning capacities 'to assess situations and make recommendations or decisions', including, most likely, and most importantly, kill decisions (Zacharias 2015).

Here too, the role of the human is that of a functional part in a wider technological assemblage of war. Moreover, where moral arguments are wielded in favour of autonomous machines as the more rational ethical actors, as Arkin seeks to make the point, the dimensions of morality in warfare shift. Where machine decisions become the normative benchmark for 'right' and 'wrong' ethical decision-making is necessarily aligned with the technological logos of algorithmic computations. By prioritising the veracity of algorithmic calculations over other relevant parameters for judgement, 'inference drawing, calculation and dispassionate rationality may all be seen as appropriate ideals' (Coker 2013: xvi). The key assertion here is the perception of *consistent* ethics as *good* ethics, as John Arquilla, Executive Director of the Information Operations Center at the Naval

Postgraduate School in the US, claims: 'my A.I. will pay more attention to the rules of engagement and create fewer ethical lapses than a human force' (Markoff 2010; Coker 2013: xviii, 178). A similar logic drives Arkin's advocacy for implementing an 'ethics module' into military robotic systems and coding IHL into it, for a more humane, machine-guided way of waging war, whereby the consistent adherence to IHL is taken as the ethical gold standard to be met. In this, ethics is conceived as a legal framework, an adherence to codes and rules, whereby unpredictable human behaviour is framed as a challenge to ethics. For Arkin, the advantage of military robots as superior ethical killing machines resides in their lacking drive for self-preservation, the absence of unchecked emotional turmoil in battle, their superior sensory capacities and information-processing powers, the absence of psychological issues that arise with humans in battle, such as scenario fulfilment. In Arkin's view, these qualities make robots ideally suited to make more ethical decisions in the heat of battle than humans. Moreover, he suggests, they should indeed serve as moral guides to humans in battle (Arkin 2010: 333–4).

The underlying question, then, shifts from whether it is ethical to kill, to whether machines would do the killing better than humans. If it has been determined by algorithmic calculation that all military-aged males in a certain geographic region, displaying certain suspicious patterns of life behaviour, pose a potential security risk, then the ethical task at hand is to kill better and more humanely. This, of course, is the logic and process that underwrites targeted killing operations conducted by the CIA and increasingly also other military and security organs. The appeal to the authority of technology shifts the concern of ethics to a purely operational space of 'killing well'. This is a space of 'necroethics', as Gregorie Chamayou puts it. 'The necroethics of the drone', he notes, 'abandons any discussion of fundamental issues', instead focusing on numbers (how many civilians killed) or *jus in bello* principles, stressing the technology's ability to better execute the laws of war without giving thought to the initial rationale for killing or indeed the altered character of war that the technology facilitates (Chamayou 2015: 162). A technology held as *a priori* more ethical becomes difficult to contest in its use, even if it is used to kill. The inadequacy narrative of the human, which I raised earlier, affords technology this authoritative position as war's most rational and ethical agent. Arkin's argument for ethical killing machines makes this starkly clear. Emotional unpredictability, messy psychology, physical limitations – all of these human qualities are threats to ethical warfare. In order to make this case, Arkin lists a litany of atrocious acts committed by soldiers during Operation 'Iraqi Freedom' (Arkin 2009a: 7). He argues that, if all soldiers were able to fulfil their soldiering role without committing atrocious acts outside of that which they were ordered to do, war would be considerably more ethical. The way forward, then, is to take the human further out of the warring process and let ethically programmed machines recommend, or indeed decide, who should be eliminated. The promise of intelligent machines is thus not simply to replace the human in war but also to guide their conduct. Here again we see the contingent, uncertain and unascertainable aspect of the human rendered in terms of risk, danger and flaw. The allegedly more ethical conduct, then, would be to equip machines with the ethical codes they need to save humans from themselves (or at least some of them).

The techno-logos is one that seeks to optimise on a numerical-digital basis. It is one of efficiency and effectiveness. For automated systems, this means consistently following an 'if-then-else' pattern, which produces consistent outputs. For intelligent autonomous weapons systems, this means that they autonomously process the sensory data input 'through optimization and verification algorithms, with a choice of action made in a fashion similar to that of humans' (Cummings 2017: 4). The decision-making process follows a probabilistic structure based on algorithmic assessment of data. And herein lies a dangerous confusion – unmanned systems are assumed to be autonomous over and above their engineered ability to function. This has important implications for the future of ethics in war.

The autonomy of robotic systems is purely a matter of engineering, of programming, of computation. Yet, once anthropomorphised, these functions are treated as equivalent to the human capacity to make decisions, to be responsible, and perhaps even to be ethical, just like a human would. The popular, but fallacious, idea of the human as a functioning machine, whose body works like a mechanism and whose brain functions like a computer,[11] fosters the anthropomorphisation of technology and the conflation of ethics with effectiveness. Take, for example, the narrative of ethical killing as advanced by Arkin. Such a narrative hinges on a notion of 'machine morality' which, when considering what technology is and does, effectively positions ethics as a programmable code (Allen and Wallach 2008). This is a fallacy, for machines need instruction through programming, and are thus tied to the moral 'input' of humans. In our ethical and political decisions, humans are social beings, 'negotiating the moral hazards and ambiguities of our human-built world' (Coker 2013: 192). Robots are not capable of this, for they are not social units. To detect whether a situation that demands an ethical decision takes place, the autonomous system would need to understand the ethical implications of any action taken. The system would have to draw on a wide variety of courses of action to assess what might be the most ethical decision in a given moment. However, if every ethical decision is one that arises in a new and different context, and each action thus has different ethical implication, such decision-making is not programmable. The fallacy of robot ethics lies in the problem that they are unable to 'recognize ethically problematic situation in the first place, or it would have to be able to think ethically on its own – and act accordingly' (Wagner 2012: 56)

Designing artificial moral agents thus relies on the *coding* of ethics and moral agency. This, in turn rests on a regulatory type of ethics, a prescription ethics and ethics as adherence to pre-established laws or rules. It is assuming that algorithms can be formulated that simulate ethical dimensions as they relate to human behaviour, better still, improve on how humans grapple with and understand ethics. The keyword here is 'Operational Morality' (Allen 2011). And herein lies the fundamental problem of ethics understood as a litany of rules and guidelines for 'correct' and 'accurate' behaviour. When ethics is understood as a matter of engineering, the notion that a unique moral demand arises with each ethical moment becomes entirely eclipsed. As a consequence, the mandate to take up the responsibility, and act morally, is obscured (Bauman 2000: 83–96). Ethics is then rendered solely as applied ethics. As Allen notes in a 2011 *New York Times* op-ed: 'the engineer might ask: Isn't ethical governance for machines just problem-solving within constraints?' Allen highlights the complexities of devising ethics for autonomous machines; however, he expresses also a frustration with the inherent uncertainty of ethical decisions. He concludes that, if the framework for ethics guidelines to be implemented into robots is left to philosophers, engineers won't ever get any 'instructions'. If it is left to engineers, ethics will be found left wanting. Anthony Beavers rightly sums up the frustration: 'Fuzzy intuitions will not do where the specifics of engineering and computational clarity are required' (Beavers 2010: 207).

This rationale of purely rational ethics assumes a guideline approach to contemporary ethics, whereby the techno-authority sets the standard, truncating all non-computational aspects of ethical decision-making as flawed. 'The problem', as Coker notes, 'is that we have a tendency these days to conclude that we should pattern all understandings in the manner of the operations of a digital data processor, the algorithmic way in which computers deal with data' (Coker 2015: 128). After all, are we not merely slow, antiquated machines, manifested in our biological and neurologically ascertainable essence? The techno-logos, however, functions on a specific setting within which this particular ethical computational logic works. This setting cannot account for compassion, empathy, surprise and broad sensory aspects of humanity that factor into each moment that might require an ethical decision. The drive towards a reliably ascertainable ethical programme must exclude that which cannot be rendered as data. An ethics that rests on such logics can never consider the uncontrollable, unascertainable, the indeterminable. It is unable to conceive of uncertainty as an inherent aspect of the ethical moment and limits our engagement with ethical content starkly and, I would argue, needlessly.

Recall that ethics-as-code is a technical practice, mimicking 'scientific analysis' and allegedly rooted in 'sound facts and hypothesis testing' (Haraway 1997: 109). Where ethics is abstracted and coded, it leaves us with little possibility to challenge the ethicality of the context within which the ethical programme unfolds. And where ethics is coded, it curbs the ethical responsibility of the individual subject. Again, I want to stress that the problem is not the existence of codes of ethics or codes as law. Understanding ethics only as code, however, attempts to normalise and homogenise something that can be neither normalised nor homogenised owing to its inherently contingent nature. This is problematic. Where the language of ethics is simultaneously the language of code, subordinated to the attainment of a specific ethical end, it may well become altogether impotent.

Conclusion

Whether robots will indeed become our technological overlords in a speculative post-human future, or whether humanity will take an entirely different turn – questions like these necessarily hinge on conjecture at this point in time. We cannot (yet) gaze into a crystal ball to ascertain the future; rather, the speculative nature of our technological future, including the use and evolution of AWS, remains, perhaps ironically, firmly situated within the realm of uncertainty. The aim of mastering uncertainty, eliminating risk, and making warfare 'safe' through increasingly autonomous technology is thus an illusion at best, a delusion at worst. Nonetheless, the drive towards more and always faster, bigger and better technology is well under way. No longer are machines the mere instruments of our will; rather, we have become subject to the logic and rationality of technology through a new human–machine relationship. This relationship betrays a complex hierarchical shift in which the human is posited as the weak link – an antiquated entity, with severe functional limitations and in need of an overhaul, or at the very least guidance from machine technologies. This is especially so in the military context. Not only do technologies shape ideas and conceptions of ethics in and of political violence; they also produce operational mandates and logics of prevention and risk elimination. The authoritative dimension of lethal unmanned and autonomous weapons, paired with a preventive security mandate, posits violent technologies as a solution to political problems. Underwritten by this bio-technological ontology and the subjectivities it produces, ethics becomes a purely technical matter to be clarified and administered by an ostensibly perfectly programmed system of infallible expertise.

This matters for ethics as it occludes any deeper engagement with what ethics asks of us. As Allen and Wallach conclude, '[t]rust and cooperation cannot be built by the dogmatic imposition of one framework over another or through the rigid application of one view of what is ethically "correct"' (Allen and Wallach 2008: 216). There is thus more to human interaction than the formal logos of mathematical computation. Truncating the non-computational aspects of humanity – the sensory, the social, the political, the historical and so on – yields a very limited idea of ethics and responsibility. 'Let's not romanticize humans', Rosa Brooks urges in her defence of robotic killing machines (Brooks 2015). However, in this particular context of uncertainty and speculation about future AI weapons capabilities, the slope is slippery. Our revaluing of machines carries an immanent devaluation of our humanity, 'for we are now trying to "moralise" weapons, an inelegant term for abdicating control over our own ethical decision-making to robots that may be better placed than us to make the right moral judgement calls' (Coker 2013: xxiii). In order to maintain a moral relationship to war as a social activity of destruction, rather than a scientific endeavour of risk elimination, and to understand what ethical responsibility is entailed in war, the science-technology narrative that works in tandem with biopolitical structures should be interrogated in current forms of warfare.

Notes

1 'And daily arise from automata, ever-better new machines. Only we remain defective, only we are obsolete' (Author's translation).
2 DeepMind is a British IT firm focusing on the development of artificial intelligence. The firm was bought by Google in 2014. Previously, Google also acquired the US-based robotics firm Boston Dynamics, a firm working closely with DARPA.
3 DARPA's Peak Soldier Performance Program was designed to artificially 'supercharge' soldiers' energy levels, alertness and metabolic capacities to allow them to operate at peak performance levels for extended periods of time (Abney, Mehlman and Lin, 2013: 7; Shachtman 2007).
4 Alex Garland's film *Ex Machina* (2015), for example explores the theme of a super-intelligent life form pitted against an intelligent computer programmer in a contest of affection to test the possibility of artificial consciousness – a Touring Test of sorts. Similarly, Spike Jonze's *Her* (2014) offers a study of the relationship between a writer and an artificially intelligent operating system,

in which the latter tragically outpaces and outgrows the very human parameters of a romantic relationship. Both films are often commented on as bell-
wethers for AI futures to come.

5   Audra Mitchell concisely summarises what broadly connects the majority of
approaches to post-humanism: 'They are linked by one common thread: the
idea that a normative, naturalized idea of the human must be challenged if
humans are to face the conditions of the universe they co-inhabit' (Mitchell
2014).

6   See for example Nick Bostrom's earlier work (2005), Anders Sandberg's and
Randal Koene's ongoing work on brain emulation as exemplary for such
techno-enthusiasm.

7   For a very nuanced and interesting discussion on human enhancement in the
military context as well as an exhaustive survey of current military projects
for the development of human enhancement technologies see Abney, Mehlman
and Lin 2013: 21–7; Galliott and Lotz 2015.

8   At the time of writing, a group of 3037 AI/Robotics researchers has signed an
open letter calling for a ban on autonomous weapons. The letter was signed by
a further 17376 endorsers, among them Stephen Hawking, Elon Musk, Steve
Wozniak and Daniel C. Dennett. The open letter, 'Autonomous Weapons:
An Open Letter from AI and Robotics Researchers', was announced in July
2015 and is available at: http://futureoflife.org/open-letter-autonomous-
weapons/.

9   Problems with AWS were first taken up by a group of scientists and scholars
who formed the International Committee for Robot Arms Control in 2009.
The issue has since gained international momentum, with a joint report by
Human Rights Watch and the Harvard Law School International Human
Rights Clinic making the first open call for an outright ban on AWS in 2012
(Human Rights Watch 2012). This report was followed by the creation in 2013
of the Campaign to Stop Killer Robots, a coalition of over thirty NGOs. The
UN Convention on Conventional Weapons has now made it a priority to
address the question of AWS in its recent meetings and expert discussions.

10 Such enthusiasm was evident, for example, in the US Armed Services Com-
mittee Hearing on 'Advancing the Science and Acceptance of Autonomy for
Future Defense Systems', held on 19 November 2015 and resonates throughout
Jon Bornstein's presentation to the NDIA Annual Science and Engineering
Technology Conference (Bornstein 2015).

11 Roger Epstein identifies this as the IT metaphor, which currently functions as
a lens to render human and computer functionality intelligible. He notes that
much of the recent enthusiasm about AI rests on this image of computers as
human brains and vice versa. For him this is a grave error, which distorts a
more profound understanding of how humans engage with their environment
and others, and consequently what the limits of AI naturally are (Epstein
2016). https://aeon.co/essays/your-brain-does-not-process-information-and-
it-is-not-a-computer.

7

Prescription drones

> Those who oppose violence with mere power will soon find that they are confronted not by men but by men's artifacts, whose inhumanity and destructive effectiveness increase in proportion to the distance separating the opponents.
>
> Hannah Arendt, *On Violence* (1970: 53)

The previous two chapters have discussed the biological-technological condition that shapes our ideas about ethics, the human and warfare. The biological-technological rationale produces subjectivities as functioning elements in a biopolitically conditioned world that is based on an organic order. Where human activities are considered primarily in their functional-technological dimension, the idea of what humans are or should become, how they should act or judge, is shaped along precisely such technological lines. The newly constituted biological-technological ecology within which we find ourselves today produces new modes of governance in all socio-political dimensions, including warfare. In this form, new military technologies represent and reinforce 'the mutual interdependence of material, biocultural and symbolic forces in the making of social and political practices' (Braidotti 2011: 329). Code – biological, technological or juridical – serves as the semantic and semiotic architecture upon which biopolitical processes unfold, and through which violent practices are sanitised. Code, as a 'structural feature of contemporary society[,] ... is both metaphor and reality' which represents and translates relations across contexts (Berry and Pawlik 2005).

The technologies touched on in the previous chapter highlight the drive towards an ever-greater and more intrusive bio-technological substance of and for new weapons technologies, and indicate the potential they have in shaping both civilian and military subjectivities. Neither one of these technologies has fully materialised yet. The production of the radically enhanced cyborg 'Super Soldier' is under way, but a fully functioning TALOS suit is a couple of years away still, and drastic biological and cognitive augmentations for humans are not yet safe or easy enough to be realised on troops at this point. Similarly, stages of autonomy in lethal weapons systems are advancing quickly, but delegating the decision to kill humans in war to an artificially intelligent machine is still some way away, for the time being at least. This leaves us no choice but to speculate or imagine what kinds of impact these potentially groundbreaking techno-logical changes might have in store for us. Conjecture, unfortunately, serves as an unsatisfying basis for a fruitful debate, and, although the discussions about the development and use of lethal AWS, AI and super soldiers are well under way, military discussions remain in the grip of a techno-optimism that sees nothing but opportunity on the horizon. There is, however, a military technology that heralds the possible impact of the unfettered push towards ever-better 'methodologies of efficient violence' and its effect on our adjustments to violence (O'Connell 2017: 144). This technology, in its lethal form, has been in use since the early 2000s and, while offering a different proposition for killing in war from an enhanced soldier or an intelligent robot, it has nonetheless paved the way for the techno-biopolitical justifications of violence discussed throughout the book. This technology is the lethal drone. The interplay of biopolitics, technology, violence, and narrowed horizons for ethical debate, let alone contestation, is patently illustrated by the current use of lethal drones by the US government (CIA and military), and its allied partners in the war on terrorism. This permits us to cast an analytical eye on the conditions, effects and impacts of a specific mode of biopolitical technology, without having to yield to speculative scenarios.

In the analysis that follows, I take drone technology to be not merely yet another tool of war. Instead, I consider the lethal drone as a system comprised of data, hardware and human 'live-ware', in which technology is embedded within a wider sociopolitical body as a form of authority. Drones, as systems, perhaps unlike any other weapons system to date, have the capacity to carve out a privileged space for legal and ethical expertise (Leander 2013: 823; Zehfuss 2011). What emerges from this is an assem-blage of techno-legal expertise that both informs and justifies the US targeted killing programme. This expertise is, I argue, a techno-biopolitical expertise, deeply infused with biological imagery and medical discourses that condition and shape pathways for ethical reasoning and the contest-ability of how violence is wielded in war. In his work on the subject, Chamayou (2015: 17) characterises the ethical framing of a lethal weapon like the drone as the discursive production of a transition in morality and value. The type of ethics produced is for Chamayou a 'necroethics', which suggests that killing can be done with care (2015: 136). This notion of killing with precision, and doing so well, is 'paradoxically vitalist', as it gives implicit priority to self-preservation. Chamayou's analysis therefore suggests a biopolitically infused epistemology in the practice of targeted killing – this is a *life-preserving* practice, albeit predominantly from the perspective of soldiers, who in turn act as defence systems of the body politic (which of course must be safeguarded at all cost). Here I want to draw attention to the prevention and prophylaxis mandates that such a 'necroethics' entails. What is at work here, I argue, is a deeper regime of techno-biopolitical expertise – an assemblage of discourses and technologies that produce and manage life on the basis of a specifically medical understanding of politics, treating the body politic as a *corpus organicus* in need of a cure.

This is the focus of this chapter and, tying together the analytical strands explored throughout the book, I argue that the use of drones for targeted killing is underwritten by technological assemblages, governed by algorithms and justified through biopolitical ethics. In drone warfare, each of these factors fits in with the others with ease in the pursuit of a sanitised, healthy and progressing humanity, and in turn produces regimes of violence that may well broaden the scope for political violence rather than limit it. Lethal drone technology and the military practices associated with it constitute the starkest indication to date of just how deeply anchored violent technologies are within contemporary biopolitics, and I highlight the semantic architecture with which claims of biological necessity position technological instruments for violence as inherently ethical. My aim, therefore, is to illustrate how influences of advanced biopolitical technologies are already at work in war and conflict today, gradually limiting the space for political and ethical contestation, while simultaneously carving out more and more space for violence-as-politics.

The chapter continues with a brief overview of the status of armed drones and their discourses on ethics today. This is not intended to be a comprehensive survey; rather, it aims to set the scene for a mapping out of the biopolitical rationale that underpins lethal drone use. Here, I focus on the production of biopolitical-technological modes of governance for a body politic as a *corpus organicus*, the health and progress of which can be ascertained through scientific and technological interventions. I then address the role of lethal drone technology, focusing on how it acquires properties of ethical expertise through its position within a broader techno-biopolitical assemblage. This assemblage functions together with medical metaphors and narratives, which advance an anthropomorphic conception of the body politic as ill and in need of targeted, expert treatment. This framing serves to occlude ethical considerations by positing lethal drones as responsible, professional tools. The techno-biopolitical assemblage thus works as described in Chapter 5, adiaphorising the practice of targeted killing by drone and pushing this practice beyond the zone of ethical contestation.

How we learned to stop worrying and love the drone

Drones are currently traded as the hottest asset in military equipment and have proliferated significantly in recent years. As a modality of war, the US has come to rely considerably on unmanned combat aircraft. At the time of writing, the US DoD reports that it operates over eleven thousand unmanned aircraft systems across domestic training events and in operations relating to the war on terrorism (US Department of Defense 2017a). For the fiscal year 2017, the budget has remained more or less constant in comparison to previous years, with a total of US$4.45 billion allocated for drone-related spending (Gettinger 2016: 1). Proportionally, the Pentagon spends most on its deadliest armed drone, the General Atomics MQ-9 Reaper drone, as well as the US Navy's unmanned MQ-4C platform, with US$1.2 billion and US$944.1 million, respectively. It also has plans to acquire 24 new MQ-9 Reaper aircraft. However, its main focus is now on improving existing craft and developing new, game-changing technologies, which are likely to include smaller, swarm-capable drones, and greater autonomy and intelligence in unmanned systems overall (Gettinger 2016: 1). Both the US Army and Navy continue to develop their respective unmanned ground vehicle and unmanned naval vehicle arsenal, making unmanned technology a crucial element in the overall US defence strategy. The Trump administration's 2018 Pentagon budget request indicates that the reliance on drone technology will continue (US Department of Defense 2017b: 81).

The majority of drones currently stocked in military arsenals around the globe are used primarily for intelligence and surveillance purposes, but the race for countries to produce their own armed drones with combat capacity is, according to widely held expert opinion, well and truly on.[1] The US administration currently uses lethal drones in two areas of engagement – in military programmes, operating in Afghanistan and Iraq as an extension of conventional warfare, and within the CIA counter-terrorism programme, which, among other activities, targets terror suspects in a number of countries with which the US is not officially at war. However, the geographies of the US drone war continues to expand into ever-widening 'battlespaces', with reports of secret US drone bases across Africa surfacing in connection with a campaign against the Islamist terror organisation Boko Haram. In this way, the drone is itself producing ever-new 'zones of war' (Shaw 2013: 553).

The US's pole position in terms of drone use and acquisition is, however, less certain today than it seemed a decade ago. The number of countries in possession of armed drones has risen to nearly thirty, and an estimated ninety countries are reported to have drones with various capacities (New America Foundation 2017; Catalano Evers et al. 2017: 2). This includes a growing number of countries that have acquired or are in the process of buying drone systems that align operationally with US platforms. The UK, France, Italy and Spain, for example, have acquired platforms that facilitate interoperability between the US and other states' unmanned technology, with a view towards greater levels of networked co-operation. This makes it all the more pressing to examine critically what practices are aligned with such interoperability, and it might be prudent to look at the US's first mover practice of targeted killing with scepticism. Other countries, particularly China, are following suit in developing both military aerial and naval drone technology, and are likely to fill a void in the market by selling their technology to states that the US has to date been restricting sales to, specifically those in the Middle East and African regions. China's counterpart to the General Atomic MQ-9 Reaper, the CH-5 Rainbow, sells for about half the price (Chen 2017). Moreover, the frequency of reports indicating that non-state actors, such as ISIS, are in possession of armed drones is increasing and it is likely that this will be a trend that is difficult to bring to a halt.

In all arenas where lethal drones are employed, it is difficult to separate the technology from the practice of targeted killing (McMahan 2013: xi). Since 2004, US-initiated drone strikes are reported to have killed between approximately 2,700 and 4,000 individuals, including American citizens, with some analysts estimating the civilian casualty rate among these statistics to be as high as 25 per cent (Asaro 2013; New America Foundation 2012; Woods and Yusufzai 2013). The majority of the deaths resulting from drone strikes occurred in 2010. The policy originated as a programme to 'capture and kill' a small number of high-value terrorist leaders in the George W. Bush years, but it has expanded its remit considerably since. Drone strikes in targeted killing missions typically fall within one of two categories: so-called 'personality strikes', where the target is known by name and deemed to be a high-value, or particularly dangerous, individual, and 'signature strikes', which target unknown persons on the basis of an algorithmic identification of life patterns. Initially, the drone programme had focused on personality strikes on targets that were well known to the intelligence community, and deemed to pose a significant terrorist threat. However, target selection by 'signature' has become an increasingly common practice, wherein 'individuals are targeted when their identities are not known but whose behaviour suggests that they are legitimate targets' (Becker and Shane 2012: Nolin, 2012). Signature strikes are underwritten by an algorithmic analysis of certain patterns of behaviour, the outcome of which is then used to construct a matrix that can distinguish between benign and malign behaviour. A highly controversial practice, President Obama had declared repeatedly that this type of target selection would be limited, if not phased out for drone strikes. While Obama at least cultivated a narrative of exercised restraint, the Trump administration has already escalated the rate of strikes to the tune of 1.6 authorised strikes a day, and an expressly less stringent care for civilian lives lost in drone operations, by shifting the strikes back into the shadows of CIA operations (Zenko 2017).

So far, we are dealing with a story of speed – of rapid advance in the development of unmanned and increasingly autonomous weapons systems, but also rapid proliferation, as more and more state and non-state actors acquire the latest in lethal technology. In contrast, the legal and particularly ethical frameworks that might limit the unrestrained use of drones advance at a somewhat sluggish pace, despite the continued and pressing need to articulate and debate the broader moral impact of the technology on all those involved with it. Whether drone technology is itself is new, or 'merely the latest iteration of a process of technological change in warfare' that has been under way for at least a century (Boyle 2015: 106), is hotly contested. Arguments offered by proponents of drones often suggest that the moral issues of drone warfare do not differ, in essence, from those posed by traditional airpower or other distanced 'stand-off' weapons (Whetham 2013: 23–4). The underlying ethos of gaining an advantage through increased distance and limited risk to troops is said to be comparable to that of a fighter jet or even helicopter. However, as Christian Enemark points out, in debates on the ethics of lethal drones arguments suggesting that there is nothing new about drones are missing the point, as, at any rate, 'they exacerbate or expand existing moral concerns about when and how force may be used' (Enemark 2014: 4). We should therefore look carefully at what logics and conditions underpin this specific technology and the practices and outcomes it produces, as well as how it might exceed, or at the least expand, moral concerns regarding the use of force. The most obvious observation about drone technology is that they have made killing easier. Lower financial and human costs, paired with an expanded reach into suspect geographies, is what makes drones such an attractive proposition for militaries – as an extended form of airpower, as well as a new technological assemblage with its own new practices. That this has made it considerably easier to meet potential threats with lethal military force is now widely accepted.

And yet current debates on drone ethics are permeated with conceptual confusions, aligning ethicality with either effectiveness or legality. As those in support of lethal drone use often posit, the suggestion is that they 'work', and by virtue of their performance – ostensibly precise, discriminatory and life-saving – they become the exemplary ethical weapon of choice (Byman 2013; Anderson 2013; Strawser 2010). In this line of reasoning, functionality, legality and effectiveness work as moral attributes by which the weapon becomes an instrument of prudence. Scholars critical of such narratives have pointed out that the idea of prudent killing is highly problematic. Maja Zehfuss, for example, explains that the conflation of precision and effectiveness with ethicality engenders a greater likelihood of warfare. The drone is emblematic in this shift toward violence as action (2011: 544). Similarly, Derek Gregory notes that the focus on legality in debates about targeted killing (with drones or otherwise) pushes both ethics and politics to the margins of debate (Gregory 2011a: 247). The conversation is further marred by the shroud of secrecy that envelops drone warfare more broadly. At the time of writing, the US administration has acknowledged only 116 civilian drone strike victims in its campaigns in Pakistan, Yemen, Somalia and Libya (Benjamin 2017). This, as commentators have widely observed, is an improbably low number. The Bureau of Investigative Journalism suggests that a realistic estimate is about six times this figure. Furthermore, details on civilians wounded, psychological damage inflicted, rationales for target selection and basic information on algorithms used for target selections are scant to say the least.

However, while there may not be a ready consensus on whether drone technology raises completely new or merely expanded moral concerns, whether effectiveness or legality constitutes ethicality in drone warfare, or indeed how many people are killed or negatively impacted by contemporary drone warfare, what is striking and unprecedented is the ethical language through which drones are styled either as 'virtuous' tools for humanitarian acts (Kennedy and Rogers 2015) or otherwise as 'legal, ethical and wise' instruments for a more ethical approach to warfare as a whole (Brennan 2012a/b; Carney 2013). As the military expert Peter Lee opines in a 2013 interview in *The Guardian*: 'if used correctly ... the MQ9 Reaper is the most potentially ethical use of airpower yet devised' (Lee 2013). Patently normative narratives like these raise questions about what is at work in the relationship between the technology, its uses and the ethical justifications given for political violence. To answer these we must look at the regime of political violence that is produced through the use of these apparently virtuous and ethical technologies. What is it that enables the framing of an instrument for surveillance and killing as an inherently ethical instrument? What sociopolitical rationale underpins this, and what facilitates the emergence of a mechanism of political violence as a form of technologically enabled 'ethical killing'? To understand this regime of ethical killing we must return to its foundations in the contemporary biopolitical condition.

Technologies of life

'Biopolitics' generally refers to institutionalised mechanisms and discourses of power over the body and biological functions at the individual and population levels, whereby political government and life government are folded in with one another for the administration of life politics (Grayson 2012: 27). In the master- and meta-mandate to secure the health, prosperity, survival and progress of a population, biopolitics is inseparably entwined with concerns and practices of control, prediction and prevention. For this, it is reliant on technologies of security, which facilitate the norms and practices which come to govern societies. As we have seen in the previous chapter, these technologies are deeply infused with a structure of biological, scientific and specifically medical thought, which facilitates the identification of normal and abnormal life, and thus the capture of that which poses a risk to life itself (Rose 2006). In an accelerated technological reality, the entwinement of material technology and biopolitics runs deep. It produces subjectivities and, in doing so, enables the emergence of new or altered modes of political violence (Braidotti 2011: 329). It also works towards understanding justifications for violence as a political mode in specific ways.

With this in mind, I argue that the rationale for positing the use of drones as an ethical means of killing is also deeply biopolitical. Drones engender the depoliticisation of human targets precisely by treating human life as a techno-political entity that can be captured in abstract, clinical terms. In effect, the human subject is rendered a data subject – a digital entity. Sophisticated visualisation techniques and technologies serve to mediate and overcome otherwise limited visual capabilities, as the drone is able to capture its subject of observation with unprecedented intimacy and distance all the same. In such robotic visualisation techniques, we find the first indication of a technological similarity of between the military and the medical. Moreover, in both domains it is an algorithmic processing of collected data that determines the parameters of risk and incursion, typically by identifying factors that fall outside of an ascertained norm. As the former CIA chief Michael Hayden freely admitted in a 2014 interview, the decision to kill is 'based on meta-data' (Cole 2014). In such a context, targeted killing practices come to reflect a logic of biopolitical power in which logistical decisions and arithmetic calculations turn political violence into a form of risk management (Grayson 2012: 28–9). It is precisely here that the rendering of human life as coterminous with an organic mechanism is relevant and enabling for violent practices of biopolitical security. Recall that, on one hand, the category 'life' appears as the target of necro-political practices *and* the terrain on which these practices play out; on the other, theories of organic life processes serve as the very basis for such practices. An expanded understanding of biopolitics that includes the politicisation of *zoe* and the *zoe*ification of politics is able to grasp this crucial dualism. Such a view recognises the structure of scientific thought, seeing the practice of analysing the body politic as resting on an understanding of biological life processes. It also ties in another, ancillary, consequence of the scientific process of truth-finding in societies where biological life dominates the social realm – namely, the possibility for a naturalistic conception of political ends, represented by the anthropomorphisation of a body politic as ill, or in mortal danger, and in need of professional intervention. Simultaneously, this body is also analysable and calculable in its functional-machinic elements, and can be intervened upon and 'fixed'. Modern bio-medical epistemologies are in this way crucial, serving to shape the production of specific biopolitical and scientific-technological subjectivities.

Nikolas Rose's work on medical life politics is instructive here. Bio-medical epistemologies understand the human in predominantly somatic terms, as an entity made up of various ascertainable and calculable life processes and patterns. Informed by bio-medical logics and technologies, such conceptions of the modern self rest on the premise that the body is 'fully intelligible and hence ... open to calculated interventions' (Rose 2006: 4). According to Rose (2006: 5), this interpenetration of science and technology yields a range of 'somatic experts', whose possession of bio-medical expertise and advanced technologies enables them to pinpoint, single out, manipulate and improve the mechanisms of life. Over time this techno-bio-medical assemblage has shaped everyday subjectivities, enjoining individuals to internalise a 'will to health' through a range of institutions, from insurance companies to health providers and government ministries. Crucially, though, for Rose this will, or drive, to health relies on 'calculations about probable futures in the present, followed by interventions into the present in order to control that potential future' (2001: 7). That which poses a risk to the *corpus organicus* must therefore be identified, as early as possible, and consequently isolated or eliminated to ensure the continued health and development of the body in question.

Here, government of life and risk management are one, and biopolitics becomes risk politics. The rise of risk thinking, however, engenders the need for risk profiling via new 'technologies of life [that] not only seek to reveal these invisible pathologies, but intervene upon them' (Rose 2006: 19). In order to prevent risks from materialising into actual threats, 'population-based calculations' are used to identify risk groups, who are then 'placed under continuing surveillance or treatment' – a form of prophylaxis that breaches, as Lorna Weir (1996: 382) points out, the 'distinction between disciplinary governance that acts on individual bodies and security governance that acts on populations'. Risk thinking thus becomes a kind of security paradigm and biopolitics a politics of security.

Rose's account of the bio-medical logics that underpin contemporary biopolitics is compelling as a lens through which to dissect the techno-biopolitical dimension of targeted killing with drones, not least as medicine more broadly 'has been central to the development of the arts of government' under modernity (Rose 2006: 28). As indicated earlier, the necropolitical 'other' of biopolitics is 'immanent within the ethos of biopolitics' (Rose 2006: 57). What are fundamental in bio-medically infused modalities of governance, however, are not questions of sovereign control but rather the bio-medical assemblage of expertise that shapes governance through risk and security thinking. Once more, technology is paramount to this development. The somatic expertise instrumental for the politics of life as such, for example, is made possible entirely through enhanced visualisation techniques for the identification and targeting of pathologies that disturb the optimisation of health. Ever-more pervasive technologies for the anticipatory analysis of life – or 'Life-Mining', as Braidotti calls it – have as their main criteria 'visibility, predictability and exportability' (2013: 61–2). This techno-bio-medical assemblage of experts and expertise shapes operations, practices and perspective on ethics in important ways.

While Rose's analysis focuses largely on the medical field of bioethics, it resonates notably in contemporary military discourse. Most visibly, the medical metaphor is often employed in the justification of strikes against terrorists, as we have seen earlier. But beyond this, principles of risk profiling in the medical field are mirrored in military strategies for identifying and managing risk in the context of targeted killing through drone strikes. The medical logic highlighted in Chapter 5 comes clearly into play here. Consider, for example, the striking alignment of language and practice in determining 'signatures' through pattern analysis, a method of target identification that is common in oncology. The logics of identifying 'signatures' through pattern analysis in medicine and 'signatures' through pattern-of-life analysis for signature strikes in drone warfare correspond here. What is crucial, then, is how the techno-biopolitical assemblage provides an ecology through which the production of practical and ethical subjectivities is shaped. The human, in a technologically driven age of biopolitics, is not simply determined by rationality but first and foremost captured in scientific terms and rendered 'predictable [and] knowable' (Berkowitz 2012). Based on probabilistic factors, identifiable character- istics and physiological or psychological knowledge linked to higher-risk categories, algorithms are conceived to identify high-risk groups and indi- viduals. For Rose this results in a hierarchical relationship that underwrites the ethics of such practices, and he harks back to Foucault's thoughts on pastoralism in linking the will-to-health subjectivity of modern humans to a pastoral power that administers the essence of risk politics. In the contemporary context, this is not a pastoralism by the state but rather 'a plural and contested field traversed by the codes pronounced by ethics committees and professional associations' (Rose 2001: 9). In short, the 'will-to-health' mandate and the resulting risk politics produce a form of ethics that is determined by the intrinsic moral 'good' implicit in health and life, and safeguarded by the various codes and law-like regulations of those that claim the scientific and technological expertise to minimise risk through effective treatments. The drone as human-machine assemblage feeds off of this, appearing to offer an appropriate means for the seemingly dispassionate and professional administration of health.

Machines of death

Drones offer a technological system that enables the collection of data, facilitates diagnostic analysis and is able to administer a course of action in specific situations of conflict with minimal risk to the operators over- seeing the use of the technology. The drone as an assemblage comprises the actual vehicle, the data collection capacities (raw data production), the algorithmic production of information (or knowledge) from raw data, the visual communications interface, the operator and the weapons charge. In this, the drone offers a platform upon which information flows, operations and the production of knowledge all adhere to scientific and computa- tional logics. This produces a combination of interlinked frames of tech- nological authority, including visual authority, algorithmic authority and interface authority, which help shape and direct moral logics at the stra- tegic level. At the tactical and operational level, the drone operator is embedded in an 'interface environment' that represents a digitalised version of specific action worlds in war. The human body here serves as a component for the hardware of the technology, which otherwise augments and replaces the capacities of the physical body. As a form of 'live-ware', the drone operators' 'eyes and operational skills [are] privileged in this assemblage' (Williams 2011: 387). This technologically shaped labour of surveillance and killing in lethal UAV operations puts unique stresses on drone operators, as they find themselves enmeshed in a human-machine assemblage that produces 'complex forms of human-machine subjectivity', which are often in tension with the cognitive, emotional and moral capacities of humans (Asaro 2013: 220).

The technological capacities of lethal unmanned systems challenge the sensory and evaluative authority of the human 'in the loop'. This is not exclusive to drone technology but is characteristic of new technological systems in war across the board. The agentic mechanism at work here fits the human into the visual and computational logics of the drone system, by which the technology offers its human operators 'superhuman' capacities. It produces an enhanced, improved, extended, sober and ostensibly neutral version of human vision. With superior sensory capacities, drones exceed human capabilities in a number of ways – from endurance (drones don't blink, neither do they need to consider pilot fatigue) to data collection and analysis (vast scopes of data can be captured and processed). This, paired with the ostensible capacity for greater precision, renders the drone not an instrument but a sanitised *guide* in the practice of killing. Set against a background where the instrument is characterised as inherently wise, the technology gives an air of dispassionate professionalism and a sense of moral certainty to the messy business of war.

The essence of this techno-authority resides in the interconnected systems of technological professionalism and morality. The drone operator is wired into a technical ethical universe; a universe that relies predominantly on scientific processes and algorithmic logics to identify correct ethical solutions. Take, for example, signature strikes. Signature strikes target unknown persons on the basis of an algorithmic identification of behavioural patterns and other markers, such as age and gender, mobile phone activity and associations. Targets for elimination are selected by way of algorithmic risk profiling on the basis of preconceived ideas of what is perceived to be a likely threat in the future. Again, the superior capacities of the drone system (surveillance, extended visuals, extensive data capture) summon the perception that patterns of normality (benign) and abnormality (malign) can be clearly identified. This is then fortified with other algorithmic data analysis programmes (SKYNET, for example), which in turn serve as a justification for the 'legitimate' killing of persons who have, quite possibly, done nothing to make themselves liable to lethal harm.

Signature strikes echo the bio-medical practice of risk profiling and surveillance with a view to prophylactic intervention. That which might pose a risk is identified and selected as a justified target merely by identifiable markers, patterns and algorithmic calculations, and in most cases the precise factors that contribute to the algorithmic determination of targets

remain opaque. What enters into the picture here, by means of surveillance and data capture technology, is the capture of life and the potential
threat to life, as a calculable and ascertainable factor. The technology itself,
understood as offering greater visual accuracy and scope of information,
invokes the supposition that clear patterns of abnormality can be detected,
which then serve to justify and legitimise a specific target selection. This
practice typifies an important techno-biopolitical dimension of drone
warfare, for the 'patterns of life' analysis employed in selecting signature
strike targets is not only a modality of cultural and spatial mapping but
also a biopolitically informed one. Joseph Pugliese explains this entwinement of algorithmic and biological knowledge in the following terms:

> The military term 'pattern of life' is inscribed with two intertwined systems
> of scientific conceptuality: algorithmic and biological. The human subject
> detected by drone's surveillance cameras is, in the first scientific schema,
> transmuted algorithmically into a patterned sequence of numerals: the
> digital code of ones and zeros. Converted into digital data coded as a 'pattern
> of life', the targeted human subject is reduced to an anonymous simulacrum
> that flickers across the screen and that can effectively be liquidated into a
> 'pattern of death' with the swivel of a joystick. Viewed through the scientific
> gaze of clinical biology, 'pattern of life' connects the drone's scanning tech-
> nologies to the discourse of an instrumentalist science, its constitutive gaze
> of objectifying detachment and its production of exterminatory violence.
> (Pugliese 2011: 943)

The distanced and disassociated modality for marking targets to be
killed, and eventually giving the kill order, produces a different commitment and engagement with acts of violence than other, less remote forms
of military engagement. It allows the human operator to avoid ethical
engagement with the visceral act of killing. Acts are virtual, and targets
are data points, mere shapes on a screen, or some other form of mediated
quasi-object in their visual interface. Captured as a data subject, or, more
abstract still, a heat signature, the human targets are reduced to the 'purely
biological categories of radiant life' (Pugliese 2011: 943). For Pugliese, it is
precisely here that the tele-technological abstraction of the act of killing
suspends 'the causal relation between the doer and the deed' (2011: 944).
I would argue, however, that it is the medical dimension that engenders
both the abstraction and the facilitation of forceful incursion via the drone
system. Specifically, the clinical gaze of bio-technological target visualisation renders the data subject a medical matter, which in turn justifies a
technological-professional response. This interlinking of the biopolitical
and the technological can clearly be discerned in the bio-logos of the lethal
drone. In target selection, the visual capacity of the drone serves as the
extended, enhanced, improved, sober and neutral eye and sight of the
human – a macroscopic device affording a clinical gaze upon potentially

problematic populations, enabling the human operators themselves to become ostensibly more sober and ethical agents (Nolin 2012). It is therefore no longer the drone that must fit into the limited and limiting visual and computational logics of the human – rather, it is the human that must fit into the logic of the machine by, for example, rotating crews to accommodate the 24-hour gaze of the drone, or by having to develop new ways of seeing (Williams 2011: 385). This is a new mapping of human and machine – a 'cyborg assemblage', as Alison Williams (2011: 384) puts it. But this assemblage is not without its hierarchies – the drone is more than a traditional prosthesis, it is indeed the *better human*. Drones are indeed designed to *outperform* the human in the tasks of war. They can remain in the skies longer than any manned aerial system, they can capture and analyse data in greater quantities, they need less frequent breaks, can perform their tasks with much greater accuracy than any human could, and they become 'smarter' with every new model and incarnation (Lin 2011; Singer 2009; 2010). Importantly, in targeted killing, the precision capacity of the drone serves as a surgical instrument, appearing to offer an ethically superior modality of necessary violence. Together, these agentic dimensions of automated lethal technology form an assemblage for the necessary and efficient elimination of bad or dangerous life.

It is here that drone technology moves from being a clinical instrument to a benchmark of professionalism and a source of moral authority. This shift rests on an anthropomorphised understanding of the drone as a peer, if not a guide, to humans and their conduct in warfare. The 'Roadmap for the Integration of Civil Remotely Piloted Aircraft Systems into the European Aviation System', published by the European RPAS Steering Group, indicates this equalisation of humans and machines in terms of the general public's expectations for ethical capacities. The report is emphatic that '[c]itizens will expect [drones] to have an ethical behaviour comparable with the human one, respecting some commonly accepted rules' (European RPAS Steering Group 2013: 44). In this formulation, the anthropomorphic logic of technology is continued as drones are posited as moral agents that can 'act' rationally, dispassionately and – at least in principle – ethically. Evident here is the techno-authoritative dimension addressed in the previous chapter. While current drone technology is primarily a technology of automation, greater levels of autonomy and intelligence in drone systems are on the immediate horizon, and this is likely to further entrench the authoritative status of technology, such that military robotics eventually becomes 'a science of imaginary technical solutions to the problem of war legitimation' (Roderick 2010: 228). In such a reality, the capacity for ethical thinking becomes purely a technical capacity, based on the assemblage of technical expertise and life data.

To be clear, the point here is neither to suggest that drones currently work without humans nor that they are indeed ethical in themselves. Rather, it is that lethal drone technologies involve the human in an ethical universe that is wired into them. This ethical universe, in turn, is one based on using scientific processes and algorithmic logics to produce 'correct' courses of action, a condition identified in Chapter 5. In the context of warfare, this is exemplified in the IF/THEN logic of current discourses on the structures of just target selections for lethal drone strikes. Bradley Strawser's defence of the ethical obligation to use drones as a weapon of choice, for example, relies on a selection of variables (X, Y, G) and principles (principle of unnecessary risk – PUR) that, combined, serve to confirm the hypothesis that using drones for killing is an ethical obligation (2010; 2013NIB: 17–18). As highlighted earlier, such forms of ethical reasoning are not unusual for analytical approaches to moral philosophy, but in this context they do represent a biopolitical and techno-scientific subjectivity in which ethical decisions are *ascertained* through data, despite the human still being formally in charge of the kill decision. In short, the techno-biopolitical assemblage of expertise in targeted killings by drones rests on a form of algorithmic governmentality, which is facilitated through the technical capacity of the drone as an agent of expertise. This assemblage enables the drone to appear as able to 'act' not only better than humans but also more ethically. This algorithmic logos, however, is also reliant on a rendering of the body politic in anthropomorphic terms, as a body in need of a cure. It is here that the professional and analytical discourse of a necessary medical procedure comes into effect.

Violence as medical necessity

The biopolitical mandate which places its political focus on the health, welfare and survival of a population gives priority to mechanisms and means of securing this goal. The violence of biopolitics thus obtains through medicalised logics of exclusion and expulsion. Techno-biopolitics, in turn, is a regime that feeds off its enemy: that which is killed produces better health. Moreover, where clinical technological modalities for preventive incursions are considered the most prudent approach to the problem of terrorism, there is a real and serious danger that the medical-military framework serves to escalate scientific-technological rationales for violence, rather than reduce violence.

The language of prudence, wisdom and lawfulness, paired with the medical dimension highlighted earlier, serves as a powerful narrative in contemporary drone wars. In recent years, and specifically under the Obama administration, the narrative spun around drone warfare has suggested that the technology possesses properties that render it morally beyond reproach. Repeatedly referred to as instruments that can eliminate cancerous terrorist cells with surgical precision, the use of lethal drones is positioned not only as legal and ethically sound but also prudent and effective (Brennan 2010, 2012a, 2012b; Carney 2013; Strawser 2013). Here Arendt's warning about organicism is worth recalling once more. Where a body politic is conceived of in organic terms, it is likely not only that artefacts determine violence's course of action but also that as the patient gets sicker, the more likely it is that the surgeon will have the last word. Not being concerned with therapeutic or remedial action, drone technology claims to deal with the worst of all pathologies by undertaking the work of the surgeon.

Only a few months after Obama publicly admitted to the existence of targeted killing operations, John Brennan (2012a) offered a medical rationale for the use of drones in his speech on *The Ethics and Efficacy of the President's Counterterrorism Strategy*. The speech highlights the positioning of drone technology as necessary and preventive medical instruments through which the 'cancer' that is Al Qaeda terrorists can be removed. The narrative of his address thus suggests not only that drones are 'wise' – because they remove risk for US personnel altogether – but also that they conform to the 'principle of humanity which requires us to use weapons that will not inflict unnecessary suffering' (Brennan 2012a). The underlying message is thus that the US possesses the wisdom and authority to perform this surgery, with 'laser-like focus', using very specific tools, and doing so responsibly. The US administration is the surgeon, countries like Pakistan, Yemen, Somalia, Afghanistan and others beset by terrorism the sickly patients. In a follow-up interview, Brennan further elaborated on the necessity of this medical intervention for the rescue and survival of humanity:

> [W]e have been very, very judicious in working with our partners to try to be surgical in terms of address those terrorist threats ... Sometimes you have to take life to save lives, and that's what we've been able to do to prevent these individual terrorists from carrying out their murderous attacks. (Brennan 2012b)

A year later still, in 2013, Brennan reiterated the notion of the US military as an expert surgeon, equipped with the right technology to heal a body politic prophylactically:

> [I]f we don't arrest the growth of Al Qaeda in a Yemen, or a Mali, or a Somalia, or whatever else, that cancer is going to overtake the body politic in the country, and then we're going to have a situation that we're not going to be able to address. (Brennan, quoted in Cherlin 2013)

The Obama White House consistently defended the use of drones as ethical and wise means to 'save American lives' (Carney 2013), and, even when issuing a caution against the gratuitous use of drones, its terminology remained within the medical register (with Obama acknowledging that drones cannot be seen as a 'cure-all for terrorism' (Obama 2013)). In order to observe professional restraint, he ensured the public that the US government has strong oversight over every strike. In so doing he only further affirmed the type of expertise upon which the lethal use of drones rests. With the current Trump administration, it is less clear how great the concern will be to frame the act of killing as a virtuous and prudent act. It is likely that President Trump will display less nuance in producing a narrative that does not jar with liberal ideas. Nonetheless, the scene has been set and the justification and alleged ethicality of the use of drones in terms of professional expertise and technological expertise are likely to serve as an undercurrent that perpetuates the practice among militaries across the globe. As highlighted in Chapter 5, the novelty of lethal drone use lies in the combination of a medical narrative to justify targeted strikes with the technological capacity to do so. The danger of this resides in the adiaphorisation of ethical matters, as elaborated earlier. This too is visible in current drone wars.

Bauman's concept of adiaphorisation captures how practical considerations can shift moral concerns into the margins. In this context, the distancing technology of the drone is central to the adiaphorisation of their use. In a series of letter exchanges with David Lyon, Bauman discusses the technological aspects of adiaphorisation, touching briefly on drones. For the most part he focuses on their surveillance capacity. However, he also notes how the ability of drone technology to facilitate swift action can circumvent complex ethical and moral considerations. As he puts it: '[T]he most seminal effect of progress in the technology of "distancing, remoteness, and automation" is the progressive and perhaps unstoppable liberation of our actions from moral constraint' (Bauman and Lyon 2013: 86). The significance of drones as a distancing technology here is thus not that they install an ethics *of* or *at* a distance. Indeed, it is not even clear that they do this – recent studies suggest that there is something strangely intimate, and traumatic, about being a drone operator (Williams 2015; Williams 2011). Rather, what the technological and scientific expertise of the drone does is move the very practice of targeted killing into a specialized and neutralized zone, a distant zone beyond ethics. In this zone 'the strikes are surgical, the passion of the killing absolutely minimal ... The casualties on our side should be nil (surgeons don't die), and the casualties on the other minimal (a few patients die)' (Mooney and Young 2005: 120). By employing cutting-edge science and technology, the act of killing can be framed as dispassionate, precise and necessary – a move that effectively neutralises it. These medical narratives, however, are based on bio-medically conditioned algorithmic subjectivities, and it is these that enable the adiaphorisation of killing (Cheney-Lippold 2011; Till 2013: 39). The process of adiaphorisation thus rests not so much on a wilful instrumentalisation of individuals (although this case could be made as well), but rather on a techno-biopolitical assemblage that frames both the individual and the body politic through forms of scientific and algorithmic expertise. The seemingly dispassionate approach and framing of targeted killing – from establishing kill zones to disposition matrices – renders the practice professional in technological and medical terms, and in so doing adiaphorises the process. Moreover, the necessarily speculative nature of the US drone programme – the inability to ascertain effectiveness and a clear body count, in terms of civilian casualties or overall damages incurred – only further empowers forms of technical expertise.

The adiaphorisation of ethical content thus rests on an assemblage of techno-biopolitical and scientific expertise that operates with and on the notion of a body politic as an organic entity. This assemblage of scientific and technological expertise then turns ethical considerations into a technical matter, neutralising these to the point of occlusion. In the practice of targeted killing with drones, not only is the targets depoliticised by being unable to defend themselves as humans rather than an algorithmically abstracted biological entity, but, as John Williams (2015: 103) notes, 'the ethical importance of the autonomous choice of the individual to engage in activity that he knows renders him potentially liable to lethal force' is also hampered. The framing of the practice of drone strikes in medical terms thus serves not only the biopolitical mandate to save lives but also the one to take them – to extinguish, as a preventive measure, any life which is deemed a danger to the survival of the population. What is an important political and ethical question (can or should lives be taken preventively as a political practice?) is effectively occluded by the implicit morality of the survival and progress mandate. Moreover, violence here is not a last resort, but a necessary option for the creation of a better, healthier political body. The wisdom or morality of this 'will to health' is difficult, if not impossible, to contest. Where scientific-technological modalities of risk management facilitate the biopolitical violence-for-health practice, the danger is that technological incursions into the bodies that make up a body politic are likely to escalate in ever-more technologically intrusive ways.

Violence by algorithm

If we take the lethal drone system to be a wise military tool, we might expect it to limit violence as part of the production of a healthy body politic. As a prudent instrument, we might expect it to contribute to an outcome towards peace, or the absence of violence. However, it is precisely here, in the technologically-mediated target selection, that the algorithmic governance of drone warfare-as-risk management finds its contradictions and limits.

To illustrate this, I return once more to the practice of signature strikes. A key task in the use of lethal drones is the identification of appropriate targets. I indicated earlier that the erstwhile criterion of targeting only 'High Value' individuals, who were known by name and position in a specific terrorist organisation (as instituted by George W. Bush in the war on terrorism) has long been supplemented with a target selection practice that sees individuals placed on kill lists, not simply because of who they are but also because of what they *appear* to be doing. In the latter instance, previously unknown and unidentified individuals are marked as potentially dangerous and placed on a kill list on the basis of their age, gender, observed patterns in their behaviour, movements and associations. It is largely unknown *what* algorithms are used or which targeting mechanisms inform this listing process. What we do know is that this process entails a combination of human intelligence (HUMINT), Social Network Analysis (SNA) and signals intelligence (SIGINT). SIGINT is the more relevant component in this context. The role of SIGINT is to feed massive amounts of data into what is known in US contexts as the 'disposition matrix', a database for tracking, capturing or killing suspected terrorists (Becker and Shane 2012). This disposition matrix is integral to imputing terrorist intent. However, the nature of the data collection process (to which the drone is instrumental) is such that, in order to yield any relevant and actionable results, the definition of 'what a terrorist is' must be quite broad. Without a single well-defined profile of a terrorist to work with, the only way that pattern-of-life analysis can identify the signature of terrorist activity is by taking a broad view of what kinds of patterns constitute a terrorist signature.

As Jutta Weber has argued convincingly, in the end it is the fear of 'failing to spot potential suspects' that provides the criteria for the algorithms which in turn determine who might be a suspect. 'Accordingly, more and more data are included in watch and kill lists', and more and more individuals are identified as potential terrorists (Weber 2016: 5). The use of algorithms to comb through the databases for patterns and associations through comprehensive recombination then yields an ever-expanding set of possible dangers and risks to be eliminated. What emerges here is a culture of searching for the Rumsfeldian 'unknown unknowns', where the aim is to find, through ever-more data points, those unknown actors posing unknown risks, rather than focusing on the empirical assessment of objective and concrete threats posed by identifiable risk actors

(Aradau and van Munster 2011). The pool of potential suspects becomes ever-greater, producing in turn what Weber calls a pre-emptive culture of techno-security, for which the drone has become emblematic. This expands, rather than collapses, time horizons for the cessation of conflict. In fact, it renders them infinite in the continued application of lethal force, with potentially ever-expanding zones of war. If the logic is to eliminate every possible risk, then there logically can never be an end to this process. In other words, there can be no end to violence.

What is evident here is a kind of strategic fallacy: an attempt to eliminate all and every possible danger using a mode of technologically informed warfare that by necessity produces ever-new categories of risk and danger. And these new categories of danger produce new conflicts, because each quantitatively produced 'terrorist' must be met with the same response as the High Value named terrorists of old. Thus, the greater the reliance on SIGINT (and the thinner the knowledge of social and cultural practices on the ground), the more unreliable this knowledge becomes as a foundation for eliminating or neutralising real threats. This matters in particular when we consider that deep grievances (such as the unjustified death of a loved one) often spawn new acts of retributive violence. Killing people on the basis of meta-data brings forth new problems of a social, cultural or political nature, which are invisible to data-driven modalities of killing.

Conclusion

The medical narrative supports the claim that outcomes through invasive violence can be ascertained. This is a symptom and consequence of the modern techno-biopolitical condition. As highlighted earlier in the book, the use of violence for the attainment of political ends is always problematic. Nonetheless, drone warfare works along these lines of reasoning. It works on the basis of an understanding that the 'cancer' of terrorism can be effectively treated. It is assumed, as John Kerry in his former capacity as Secretary of State suggested, that 'the program will end as we have eliminated most of the threat and continue to eliminate it' (Golovina and Wroughton 2013).[2] This can be positioned as a prudent course of action, because the technological systems that allow for a medicalised intervention are sophisticated enough to shift warfare into a space of war*care* instead. Thus, what we are seeing is the production of a new type of ethical proposition – a 'necroethics' (Chamayou 2015: 136) – that paradoxically frames killing as a humanitarian, moral act of care. In this universe, the act of ethical killing is biopolitically justified for purposes of prophylaxis and prevention, in order to maintain the homeostasis of an organic entity. The medical narrative that works across military-medical spaces manifests a mandate to let the professionals deal with problems in an efficient and effective manner. After all, it is they who have the knowledge of pathological afflictions, and it is they who have the tools needed to address them. In this way, drone technology provides both the medium and the expertise needed to undertake targeted killing with a professional ethos and neutral distance, just as a physician would. What emerges in this context is not a form of ethics at all, but rather a narrowing of ethical horizons altogether in favour of swift action posed as vital and necessary. This constitutes an adiaphorisation of drone warfare, mediated through a scientific structure of biopolitical thought and technological expertise, which is so clearly posited as ethical that it becomes difficult to contest. And where violent acts cannot be contested on ethical grounds, limits to violence recede.

Notes

1   For example see Lorenz, Mittelstaedt and Schmitz (2011), Ouden and Zwijnenburg (2011) and Sanger (2011).
2   This is reflected, for example, in a statement that the US Secretary of State made in July 2013 to Pakistan's Prime Minister Nawaz Sharif in the context of the controversial strikes in Pakistan. Kerry states in an interview that 'the program will end as we have eliminated most of the threat and continue to eliminate it' (Golovina and Wroughton 2013).

Coda: Ethics beyond technics

The voice of individual moral conscience is best heard in the tumult of
political and social discord
Zygmunt Bauman, *Modernity and the Holocaust* (2012: 166)

In this book I have explored how the techno-biopolitical entanglement of
our present shapes our adjustment to violence as a political solution,
impacts on our understanding of ethics and alters our ability to contest
ethically relevant decisions in warfare. The infusion of a biopolitical
rationality into the idea of the human is as relevant in our contemporary
present as it was in the context of Nazi Germany. Today, this biopolitical
rationality is deeply interwoven with an environment of accelerated technological substitutions for an ever-growing realm of human activities. At
stake in this present is nothing less than the production of 'who we are,
how we view the world, how we interact with each other'; in short, at stake
is nothing other than the shaping of human subjectivity and the ethical
and political conceptions that emerge therefrom (Hardt and Negri 2006:
66). The biopolitical rationale and the technologically mediated ecology in
which humans today find themselves produces a subjectivity that stands
under the imperative to eliminate risk, mitigate uncertainty and, most
importantly, produce predictable, certain outcomes. It is a data-driven,
algorithmically oriented subjectivity, the logic of which requires that alterity and difference must be captured in homogenising terms to render the
inherent unpredictability and uncertainty of aleatory life mitigated by the
calculability of common factors.

Perhaps paradoxically, the framing of alterity and unpredictability as a
problem for humanity, which is to be politically addressed, produces a
potentially endless number of new categories of risk and potential dangers
for consideration. A scientific mindset assists in 'securing' the abnormali-
ties that are to be addressed; a technological orientation facilitates the
processes as a panacea to the problem of unruliness. With ever-greater
scopes of collecting calculable features about life – its data – the fiction

manifests that eventually there will be so much data about life that every political problem can be knowable. And with progressive advancements in producing machines and systems that mirror our own datafied understanding of the human body and brain, a technological universe is created in which we have created machine-gods in our image, which then shape humans in their image; a kind of solipsistic religion of technological salvation. I have argued here that the focus on, and faith in, scientific-technological solutions harbour the potential to promote rather than limit violence within a biopolitical context. America's drone war against terrorism is a case in point.

Technological blinkers

A sober assessment of the outcome of over a decade of lethal drone use in the war on terrorism points towards a problematic development related to the speed and scope with which lethal drone technology (and its related practices) proliferates. Possession and use of lethal drone technology is no longer the exclusive domain of the US. As mentioned earlier, many states and non-state actors are following suit in discovering the technology as useful for a low-risk strategy to address a problem with force. In this, norms that would require the strict adherence to political accountability and transparency for acts of violence are slipping. The United Kingdom is a case in point in evoking executive powers in drone strikes since 2015 and the attorney general Jeremy Wright's 2017 statement that the government's lethal strike policy (drones or otherwise) would no longer require specific evidence of terrorist acts for a basis to pre-emptively target individuals, rather, in line with US policy, it would align its policy with 'the principles [of pre-emptive self-defence] that the US have already adopted, which [the UK] think strikes the right balance of all the different factors' (Wright quoted in Bowcott 2017). The US principles of pre-emptive self-defence are, however, notoriously elastic. As more countries acquire drone technologies that align operationally with US capabilities, the pressure to adopt a lethal strike policy grows. Here, first-mover practices and narratives set precedents for the normalisation of killing off suspected terrorists. Moreover, a more scaled down, but nonetheless effective version of drone technology proliferates among non-state actors including ISIS, as terrorist organisations use commercially available drones and modify them to suit their own violent goals, often loading them with explosives and sending them to their intended targets via remote control. As more sophisticated drone technology becomes available more widely commercially, this adaptation to, and adoption of, drone technology by terrorist groups is likely to become a more serious problem, specifically as terrorist groups are likely to be less squeamish about equipping drone technology with chemical, biological or nuclear capacity. But it is not only the proliferation of the technology that produces new practices of violence among terrorist populations. It is also the dialectic interplay between new technologies and their logical counterparts. Akbar Ahmed, for example, argues that the contemporary practice of suicide bombing among Waziristani tribesmen emerges directly as a response to drone operations employed after 9/11 (Ahmed 2013: 89). Rather than eliminating the terrorist threat with the production and use of drones, a new, dangerous dimension of drone use – a new category of risk – is produced.

This counterproductive dimension of drone war logic is perhaps best viewed with the military-medical lens addressed in Chapter 5 in mind. The technologised military gaze, like the scientific medical gaze, 'tells us as much about the performance of healing, suffering and dying, as chemical analysis tells us about the aesthetic value of pottery' (Illich 1976: 103). In short, the scientific-technological approach to both medicine and military problem-solving may well lose sight entirely of social and cultural contexts which have relevance for political problems. In the context of medicine, the counterproductive effect that results from an increased reliance on scientific-technological solutions to all medical problems is known as *iatrogenesis*. Derived from the Greek words *iatros* (physician) and *genesis* (origin), the term denotes a condition 'resulting from the activity of a health care provider or institution; said of any adverse condition in a patient from treatment by a physician, nurse or allied health professional' (Illich 1976). The philosopher Ivan Illich engages this concept to critique an overreliance on financially expensive technologies of intervention in the pursuit of healing. In his assessment, it is the 'increased investment in medical technologies' that yields counterproductive outcomes, creating ever-growing populations in peril and with a diminished autonomy to address their conditions (Illich 1975: 78).

Called into question here is the autonomy of those subject to regimes of scientific-technological incursions. Here too, the dimensions and impact of technology are greater than their efficacy as tools for a specific job; rather they constitute an assemblage in which materials, humans, environments and practices form a network from which new, sometimes unintended practices and outcomes emerge. This perspective is easily transferred to the discussions presented in this book. Where autonomy is compromised, the possibility for moral autonomy is called into question. Where the technological authority of killing machines meets with a biopolitical rationale of eliminating that which stands in the way of the life processes and progress of a political community, the worry is that an ethical vacuum emerges in which the choice to kill is the rational choice, and therefore the first, not the last, resort. This is precisely what the push for artificially intelligent autonomous technologies heralds. The questions at the forefront of current debates on autonomous lethal weapons systems focus on accountability and responsibility when machines kill in an unintended manner. They centre on what 'meaningful human control' denotes in operating such systems (Moyes and Roff 2016). These are important questions, specifically as the development of autonomous intelligent weapons systems advances rapidly. However, they overlook other, pressing questions about the production of new practices that require ethical consideration beyond the killing of people, and they neglect to acknowledge the degree to which we have already acquiesced to the techno-biopolitical logic of eliminating risk (the life of others) through technologically authoritative systems. The question should perhaps be: what new ethical responsibilities arise from new killing machines?

The question of agency looms large. Theories of materialism posit that we are always already assemblages of materials and substances, and are always dependent 'on the collaboration, cooperation or interactive interference of many bodies and forces' (Bennett 2010: 21). In this, technology plays an important role as having the capacity to 'authorise, make possible, encourage, make available, allow, suggest, influence, hinder, prohibit …' certain acts, or the possibility to recognise certain acts (Latour 2005: 104). We then always find ourselves in a web of relations for our political and ethical acts. This understanding of human embeddedness and interrelationality within the world and its subjects and objects is notable in Arendt's account of the human condition in modernity. But unlike materialist theories, she ascribes the possibility for political agency clearly to the human. In this, the human has, in theory, the capacity to make new political beginnings; this possibility is hampered only when the space and horizon for political action proper are eclipsed.

The explorations in this book have sought to highlight how human agency is hampered where mere behaviour cannot be transcended and action becomes impossible. Plurality, freedom and speech are crucial and fundamental aspects in the Arendtian account of politics. In a highly technologised biopolitical context, all three are altered to the point of irrelevance. In order to act politically, elements of contingency, risk and unknowability must be embraced for Arendt, as political outcomes themselves can only ever be characterised by uncertainty. Such is the human condition. But it is precisely in this that the potentiality for new beginnings and the possibility to overcome 'dark times' in humanity lies. Arendt considers this 'dark side' of human affairs, but does not bemoan it, rather she considers it part of the human condition in all its positive potentialities. What she does bemoan is the loss of human action and politics in a biopolitically technologised modernity, as 'to try to "make" human life by eliminating its unpredictability is to destroy the human condition' (Parekh 2008: 172). In Arendt's own words: 'the fact that contemporary politics is concerned with the naked existence of us all is itself the clearest sign of the disastrous state in which the world finds itself – a disaster that, along with all the rest, threatens to rid the world of politics' (Arendt 2006b: xx). When this leads to a diminishing of our ability to take responsibility for acts of violence, or, moreover, shifts the parameters for what we have a responsibility towards, then our relation to technology becomes a political question as much as an ethical question of the highest order. Where political agency is limited, ethical agency is equally hampered.

Precarious ethics

In a technology-dominated and biopolitically informed world in which the drive towards post-human desires – in which machines might indeed 'realize their potential as the masters of man' – is becoming increasingly visible, what can we then make of our responsibilities in the context of political violence? The question of violence, and specifically political violence is a quintessential ethical question. It relates directly, and in the most radical sense, to the responsibility we have, as humans to other humans. As Michael Dillon notes: 'the question of the human is not only central once again, but also is that of the *understanding* of politics. The one is always intimately related to the other' (Dillon 1998: 545, emphasis added). As a growing part of our contemporary world is dominated by technological substitutes for the human, a growing remit of 'our ethical and political thinking is tested through technologically informed rational choice models' (Berkowitz 2010). This is the logic I have sought to highlight in Chapters 5 and 6 as it forms the centre of discussions on technologically mediated modalities of political violence, whereby the 'moral demand is transferred from the human to the machine' (Babich 2012: 28; Anders 2010: 40). This is underwritten by a technology fetish that demands that whatever is technically possible must be realised without delay (Alston, Rona and Hakimi 2012; Babich 2012: 28). In this, an understanding of ethics as a comprehensive demand towards others, arising from non-computational, plural contexts of experiences, socialisation and sensory dimensions, is pushed to the margins of what it means to act ethically. By searching to prescribe ethical principles to an abstracted set of instances and occurrences, the contingent character of the encounter with the other, to speak in Emmanuel Levinas's terms, is disregarded in the assumption that ethical dilemmas can be resolved (Hutchings 2010: 152). When we understand ethics in the Levinas–Derridean tradition as being 'centered on the relationship between the subject and otherness in the mode of indebtedness, vulnerability, and mourning', and as arising from the encounter with the other, which does 'neither threaten to punish or promise a reward' but simply triggers, by sheer presence, an ethical moment, it is precisely in this very moment of the ethical decision that the actual indeterminacy of ethics lies (Bauman 2012: 214; Braidotti 2011: 302; Dauphinee 2009: 243). Similarly, for Derrida, it is in precisely this moment of an ethical demand being made that requires a decision that ethics unfolds (Raffoul 2008: 273). It is the very impossibility of knowing the 'right decision' that makes ethics possible and that requires that ethics is considered not in a universalised and universalising abstract set of rules but with each encounter. Derrida elaborates the rationale for this complex distinction between ethics as principles and ethics as reliant on the very possibility of the impossible as follows:

> If I know what I must do, I do not take a decision, I apply knowledge, I unfold a program. For there to be a decision, I must not know what to do ... The moment of decision, the ethical moment, if you will, is independent from knowledge. It is when I don't know the right rule that the ethical question arises. (Derrida 2004, cited in Raffoul 2008: 285)

In other words, the ethical decision, contrary to modern aspirations of applying ethical principles as ethical laws, arises from a status of non-knowledge. This status of non-knowledge is a precarious state, it risks failure, it requires courage. It is the encounter with the other that bestows not only an implicit vulnerability on the other but also a vulnerability of the self in being unable to know, to have certainty. For such ethical decision to occur authentically, one must thus accept the very fallibility of the human. Right and wrong can therefore not be secured, which, in turn, means life cannot be definitively secured. This unsettles the techno-biopolitical rationale entirely. The potential 'fallibility' of the decision-maker in each ethical decision is in tension with the perception of the human as controllable, calculable and utilisable entity within project humankind. It is only in the biopolitical context of the human understood as a calculable and mathematisable being that the notion of 'fallibility', 'failure' and 'error' of the human can emerge in the first place, and ethics and moral acts can be framed in terms of 'correct' and 'incorrect'.

This tension in the perspective of ethics is evident in the apparent disconnect between ethics as understood as the moment of responsibility for others on one hand, and professional, technical ethics on the other, where the notion of improving moral standards is a matter of providing an enforceable ethical code and the failure of adherence to stipulated ethical behaviour is 'blamed on the faults of the ethical code or the laxity of the organs of its promotion and enforcement' (Bauman 2000: 86). The logical solution then is a more stringent ethical code, or more rigorous enforcement. Rather than shaping the human towards greater ethical sensibilities, ethical responsibility is externalised. The ethical code is therefore performative as a technical and professional 'moral compass to regulate behaviour' (Balmer and Rappert 2007: 58). However, it cannot inform action proper. Where ethics is coded, it contains the illusory promise of certainty, the fallacy of being able to offer a technical way of resolving ethical questions. The actual ethical question is thus obscured. The fixation with ethics codes, ethical principles and ethical laws produces frameworks in which blind-spots for moral concerns arise in the interstices of the frameworks; it pushes some morally relevant aspects to the margins of the debate, while zooming in on others. In the case of drone warfare, the preoccupation with lives saved (of military personnel) and lives taken (as collateral in the war on terrorism) obscures the production of wounded and maimed bodies, psychological harm inflicted or disruptions and impediments of political communities in targeted populations. When code is paramount, that which is pushed to the margins of code can thus find no form of expression and consequently no way of addressing adequately.

If we want to rethink ethics, we ought to consider it in the context of politics, the human and the complexities of today's technologised world. As Braidotti suggests, 'an ethics worthy of the complexities of our time requires a fundamental redefinition of our understanding of the subject in his location' (Braidotti 2011: 300). In this, ethics ought to be able to incorporate the uncertain, the unascertainable, the indeterminable implicit in multitudes of messy beings. Braidotti posits that in a technologically fast-paced and highly globally diverse world of non-unitary subjects, ethics must be reconsidered as a 'matter of experimentation, not of control by social techniques of alienation' (2011: 348). I thus suggest, with Braidotti, that ethics is action, is doing, rather than a theory or a category for application. In her study of *Ethics and Politics After Poststructuralism* Madeleine Fagan argues that

> the 'ethical' should not be understood as a label; it does not mean 'good' or 'right'. Likewise it is not an evaluation or guide. Rather, both the ethical and the political are descriptions of the context in which we find ourselves; compelling and irreconcilable obligations can and do happen in a forceful way, without foundations. (2013: 8)

Understood as an action, ethics becomes much more closely tied with politics as action and the two share essential aspects. Here, I suggest we follow Braidotti in attempting to reinsert 'the "active" back into activism as an ethical as well as political project' (2011: 348), and Fagan in making the link between ethics and politics as action. Ethics is action, which cannot be secured.

The political substance of ethics

The ethicality of ethics implies fallibility, uncertainty and contingency, and only through embracing risk, in political and ethical acts, can responsibility for actions toward the other be taken. As indicated in Chapter 5, the point here is not the outright rejection of code: on the contrary, coded guidelines, regulatory frameworks of law are of immense significance and value in a sociopolitical context and they provide the stability that supports the political body. But what is at stake here is whether we should understand ethics as code, and conceive of ethics as a problem than can optimally be solved by the application of regulations. To posit that ethics can be ascertained is, indeed, an illusion and one that limits the responsibility of humans to be morally accountable for acts that are committed in their name, in the name of humanity. It is here that Arendt's warning holds sway that with the subsuming of speech into the expressions of symbolic and abstract scientific codes it becomes difficult to break away from the anti-political boundaries of behaviour. Yet Arendtian subjects have the capacity for action, if they decide to take on a risk and engage in unpredictable outcomes – engage in political action proper.

Here, the issue of politics proper becomes so salient for ethics. Where politics is understood as management and administration, an engagement with ethics as outlined above is relegated to the margins, subsumed by the professional authority of technological expertise and instruction. Melissa Orlie elaborates the relationship between ethics and politics convincingly: 'to be free is to act unpredictably, to upset expectations based on what you appear to be in order to reveal who you are becoming. To be incalculable is to act responsibly because we thereby abate the deadening weight of harm and wounds' (Orlie 1997: 197). For Orlie, the meaning and significance of ethical political action is precisely to be found in the fact that, even under totalitarian oppression, most people comply with immoral rules – but not all. Each disruption of coded and rule-based political acts that might, by an individual or collective, be considered unethical cements the human capacity to interrupt 'social rules' and act truly political, free and responsible. Orlie states: 'Paradoxically, responsibility demands incalculability and unpredictability, while freedom requires that we be responsive to the harms that invariably accompany the good we would do. In short, to live ethically, we must think and act politically' (Orlie 1997: 169–70). It is in the 'unobvious decision' within which ethics is revealed (Coker 2013: 175). It is by doing what is not expected, and which might pose a certain, further risk, in which ethics, as a form of action, is actualised.

Simon Critchley highlights this possibility for a rethinking of ethics by conducting a thought experiment for an alternative outcome in the response to the World Trade Center attacks in September 2001. It is worth quoting at length:

> Ask yourself: what if nothing had happened after 9/11? No revenge, no retribution, no failed surgical strikes on the Afghanistan-Pakistan border, no poorly planned bloody fiasco in Iraq, no surges and no insurgencies to surge against; nothing. What if the government had simply decided to turn the other cheek and forgive those who sought to attack it, not seven times, but seventy times seven? What if the grief and mourning that followed 9/11 were allowed to foster a nonviolent ethics of compassion rather than a violent politics of revenge and retribution? What if the crime of the Sept. 11 attacks had led not to an unending war on terror, but the cultivation of a practice of peace – a difficult, fraught and ever-compromised endeavor, but perhaps worth the attempt? (Critchley 2011)

Critchley considers here the unobvious decision, perhaps a decision that might be the more ethical decision in assessing the human impact and experience of warfare. He sees the moment of forgiveness as an ethical response. A moment that brings with it radical uncertainty and vulnerability. And it is precisely this moment of forgiving that Arendt sees as an authentic act of creating new beginnings in human relations (Canovan 1995: 153; Young-Bruehl 2006); forgiving as an 'alternative to revenge' which has the capacity to interrupt the 'chain of automatic action and reaction in which human affairs so easily become trapped' (Canovan 1995: 191). And it is this moment of forgiving that is properly political in Arendt's terms.

When we encounter discourses that seek to normalise drone strikes, that claim the moral superiority of lethal military robotics or that advocate practices of extraordinary rendition and torture under the mandate of the security of humanity, we ought to question the subjectivities that rive rise to a politics and ethics that facilitate such discourses and narratives. As our 'ethical imagination is failing to catch up with the fast expanding realm of our ethical responsibilities', technological subjectivities and materialities may well remove us ever further from 'the responsibility we owe to our fellow human beings' (Coker 2008: 151) – especially in situations of war and armed conflict. Further research into the ethical underpinnings of a fully technologised conception of warfare and armed interventions are needed urgently, as new technologies develop rapidly.

Conclusion

When Arendt wrote *The Human Condition* in 1958, she could not quite have foreseen the extent to which humans and their capacity for politics might become imperilled in contemporary modernity. But, like Marcuse and Anders, she had an inkling as to the implication this development of technology as a 'biological process' might have for a society and political community, and presciently attested to our capacity and desire to exchange 'given' life 'for something he has made himself' (Arendt 1998: 3). The question that stood at the heart of Arendt's inquiry into the human condition, and which has also inspired the inquiries of this book, is an open question. It has no definitive answer and requires to be posed anew in each different social and political context. For this reason alone, it bears repeating: *What are we doing?* To pose this question creates an opening of the category 'ethics' and its relationship to politics today. This opening, in turn, allows for a rethinking of how we can act ethically and politically in an increasingly complex and technologised international world. Where human agency and autonomy are eclipsed by machine thinking and acting, we may no longer be able to offer an answer to this paramount question. Avoiding such a condition requires a political and an ethical effort.

The inquiry comprised in this book is limited to a critique. What it cannot offer, at this stage, are practical solutions and concrete ideas as to how to reform just war discussions, practicalities of guidelines for fighting in war or concrete reforms for policies as to what, indeed, constitutes ethical acting in international violent engagements. This falls within the remit of further and much needed research, especially as the development of new, sophisticated, 'smart' technologies races ahead and towards capacities that we have yet to comprehend. What this book has hoped to achieve, however, is to provide an insight into how our subjectivities are shaped in specific ways that change our engagement with ethics in the political context; how we are, indeed, biopolitically conditioned, and how this, in turn, conditions our engagement with one another. There is much to be done to overcome the biopolitically informed limitations highlighted in this book, and it requires first and foremost an interrogation of the subjectivities with which we shape the world in which we live. This book and the critical analysis it offers hopes to aid in creating the opening needed to freshly consider our actions and interactions in a shared world

in an Arendtian sense, with thought and care, ethically, and politically
ever anew.

Notes

1 This refers to the title of a panel – 'Will Machines Realize Their Potential as
  the Masters of Man' – which was held as part of the Being Human in an
  Inhuman Age Conference held at the Hannah Arendt Centre, Bard College,
  New York, in October 2010.

References

Abney K., Bekey G. and Lin P. (2008) *Autonomous Military Robotics: Risks, Ethics and Design*, Prepared for US Department of Navy, Office of Naval Research, California Polytechnic State University, http://ethics.calpoly.edu/ONR_report.pdf.

Abney K., Bekey G. and Lin P. (2011) *Robot Ethics: The Ethical and Social Implications of Robotics*, Cambridge, MA: MIT Press.

Abney K., Mehlman M. and Lin P. (2013) *Enhanced Warfighters: Risk, Ethics and Policy*, New York: The Greenwall Foundation.

Ackerman E. (2015) 'We Should Not Ban "Killer Robots" and Here's Why', *IEEE Spectrum*, http://spectrum.ieee.org/automaton/robotics/artificial-intelligence/we-should-not-ban-killer-robots.

Agamben G. (1998) *Homo Sacer – Sovereign Power and Bare Life*, Stanford: Stanford University Press.

Agamben G. (2000) *Means Without Ends*, Minneapolis: University of Minnesota Press.

Agar N. (2010). *Humanity's End: Why We Should Reject Radical Enhancement*, Cambridge, MA: MIT Press.

Ahmed A. (2013) *The Thistle and the Drone: How America's War on Terror Became a Global War on Tribal Islam*, Washington, DC: Brookings Institution Press.

Allen A. (2000) 'The Anti-Subjective Hypothesis: Michel Foucault and the Death of the Subject', *The Philosophical Forum* 31(2): 113–30.

Allen A. (2002) 'Power, Subjectivity and Agency: Between Arendt and Foucault', *International Journal of Philosophical Studies* 10(2): 131–49.

Allen C. (2011) 'The Future of Moral Machines', *New York Times*, 25 December, http://opinionator.blogs.nytimes.com/2011/12/25/the-future-of-moral-machines.

Allen C. and Wallach W. (2008) *Moral Machines: Teaching Robots Right from Wrong*, Oxford: Oxford University Press.

Alston P., Rona G. and Hakimi M. (2012) 'Live Seminar 46: Emerging Challenges in the Age of Drones: Target Killings and Humanitarian Protection', *Humanitarian Policy and Conflict Research (HPCR)*, Harvard University, https://vimeo.com/50703824.

Amadi O. and Wonah E. (2016) 'The Organic Theory of State and the Philosophical Tradition: the Case of Plato and Aristotle', *International Journal of Science Inventions Today* 5(5): 415–22.

Anders G. (2010) *Die Antiquiertheit des Menschen – Ueber die Seele im Zeitalter der zweiten industriellen Revolution*, Munich: C.H. Beck.

Anderson B. (2011) 'Facing the Future Enemy US Counterinsurgency Doctrine and the Pre-insurgent', *Theory, Culture and Society* 28(7–8): 216–40.

Anderson K. (2013) 'Drones Are Ethical and Effective', *Oxford Union Debate*, Oxford University, www.youtube.com/watch?v=G99A8LAKnaU.

Ansorge J. (2012) 'Orientalism in the Machine', in Barkawi T. and Stanski K. (eds), *Orientalism and War*, New York: Columbia University Press: 129–50.

Aradau C. and van Munster R. (2011) *Politics of Catastrophe: Genealogies of the Unknown*, New York and London: Routledge.

Arendt H. (1953) 'Ideology and Terror: A Novel Form of Government', *The Review of Politics* 15(3): 303 –27.

Arendt H. (1967) 'Legitimacy of Violence. Theatre for Ideas', Hannah Arendt Papers at the Library of Congress, Library of Congress Archive, New York.

Arendt H. (1970) *On Violence*, Orlando: Harcourt, Brace and Company.

Arendt H. (1998) *The Human Condition*, Chicago: University of Chicago Press.

Arendt H. (2002) 'Karl Marx and the Tradition of Western Political Thought', *Social Research* 69(2): 273–319.

Arendt H. (2003a) *Responsibility and Judgment*. Ed. Kohn J. New York: Schocken Books.

Arendt H. (2003b) 'What Is Freedom', in Baehr P. (ed.), *The Portable Hannah Arendt*, London: Penguin Books.

Arendt H. (2004) *Origins of Totalitarianism*, New York: Random House / Schocken Books.

Arendt H. (2005) *The Promise of Politics*. Ed. Kohn J. New York: Schocken Books.

Arendt H. (2006a) *On Revolution*, London: Penguin Books.

Arendt H. (2006b) *Between Past and Future*. Ed. Kohn J. London: Penguin Books.

Arkin R. (2009a) 'Ethical Robots in Warfare'. *IEEE Technology and Society Magazine* 28(1): 30–3.

Arkin R. (2009b) *Governing Lethal Behavior in Autonomous Robots*. Boca Raton: CRC Press.

Arkin R. (2010) 'The Case for Ethical Autonomy in Unmanned Systems', *Journal of Military Ethics* 9(4): 332–41.

Arkin R. (2015) 'Lethal Autonomous Weapons Systems and the Plight of the Noncombatant', Geneva, www.unog.ch/80256EDD006B8954/(httpAssets)/FD01CB0025020DDFC1257CD70060EA38/$file/Arkin_LAWS_technical_2014.pdf.

Asaro P. (2012) 'On Banning Autonomous Weapon Systems: Human Rights, Automation, and the Dehumanization of Lethal Decision-Making', *International Review of the Red Cross – Humanitarian Debate: Law, Policy, Action* 94(886), www.icrc.org/eng/resources/international-review/review-886-new-technologies-warfare/review-886-all.pdf.

Asaro P. (2013) 'The Labor of Surveillance and Bureaucratized Killing: New Subjectivities of Military Drone Operators', *Social Semiotics* 23(2): 196–224.

Babich B. (2012) 'Nietzsche's Post-Human Imperative: On the "All-too-Human" Dream of Transhumanism', *The Agonist* 4(2): 1–39.

Bachmann S. and Haeussler U. (2011) 'Targeted Killing as a Means of Asymmetric Warfare: A Provocative View and Invitation to Debate', *Law, Crime and History* 1: 9–15.

Bäck A. (2009) 'Thinking Clearly about Violence', in Bufacchi V. (ed.), *Violence, A Philosophical Anthology*, Basingstoke: Palgrave Macmillan.

Bacon F. (1612) 'Of Friendship', www.authorama.com/essays-of-francis-bacon-27.html.

Bajema C.J. (1988) 'Charles Darwin on Man in the First Edition of the Origin of Species', *Journal of the History of Biology* 21(2): 403–10.

Balibar E. (2007) '(De)Constructing the Human and Human Institution: A Reflection on the Coherence of Hannah Arendt's Practical Philosophy', *Social Research* 74(3): 727–38.

Balmer B. and Rappert B. (2007) 'Rethinking "Secrecy" and "Disclosure": What Science and Technology Studies Can Offer Attempts to Govern WMD Threats', in Balmer, B. and Rappert, B. (eds), *Technology and Security – Governing Threats in the New Millennium*, New Security Challenges, Basingstoke: Palgrave Macmillan: 45–65.

Barder A. and Debrix F. (2009) 'Nothing to Fear but Fear: Governmentality and the Biopolitical Production of Terror', *International Political Sociology* 3: 398–413.

Barkan S.E. and Snowden L.L. (2000) *Collective Violence*, Welwyn Garden City: Pearson Higher Education.

Bassett C. (2013) *The Philosopher, the Socialite, the Engineers, and the Spy: 'Cyber-cultural' Debates in 1964*, London: King's College London.

Baudrillard J. (2001) *Jean Baudrillard, Selected Writings*, 2nd ed. Ed. Poster M. Cambridge: Polity.

Baudrillard J. (2009) *The Transparency of Evil – Essays on Extreme Phenomena*, London: Verso.

Bauman Z. (1993) *Postmodern Ethics*, Oxford: Blackwell.

Bauman Z. (2000) 'Ethics of Individuals', *Canadian Journal of Sociology* 25(1): 83–96.

Bauman Z. (2007) *Liquid Life*, Cambridge: Polity Press.

Bauman Z. (2012) *Modernity and the Holocaust.* Cambridge: Polity Press.

Bauman Z. and Lyon D. (2013) *Liquid Surveillance*, Cambridge: Polity.

Bayertz K. (2002) 'Self-Enlightenment of Applied Ethics', in Chadwick R.F. and Schroeder, D. (eds), *Applied Ethics – Critical Concepts in Philosophy*, Vol 1, London: Routledge: 36–52.

Bazzicalupo L. (2006) 'The Ambivalences of Biopolitics', *Diacritics* 36(2): 109–16.

Beauchamp T. (1984) 'On Eliminating the Distinction between Applied Ethics and Ethical Theory', *The Monist* 67(4): 514–30.

Beauchamp T. (2005) 'The Nature of Applied Ethics', in Frey R.G. and Wellman C.H. (eds), *A Companion to Applied Ethics*, Malden: Blackwell: 1–16.

Beavers A.F. (2010) 'Editorial', *Ethics and Information Technology* 12(3): 207–8.

Becker J. and Shane S. (2012) 'Secret "Kill List" Tests Obama's Principles', *The New York Times*, 29 May, www.nytimes.com/2012/05/29/world/obamas-leadership-in-war-on-al-qaeda.html.

Bell C. (2012) 'Hybrid Warfare and Its Metaphors', *Humanity: An International Journal of Human Rights, Humanitarianism and Development* 3(2): 225–47.

Bell D. (ed.) (2010) *Ethics and World Politics*, New York: Oxford University Press.

Belvedere M. (2015) 'Donald Trump: We Should Attack Terrorist Oil, Bank Resources', *CNBC*, 16 November, www.cnbc.com/2015/11/16/donald-trump-we-should-attack-terrorist-oil-bank-resources.html.

Benbaji Y. (2008) 'A Defense of the Traditional War', *Ethics* 118: 464–95.

Benhabib S. (1996) *The Reluctant Modernism of Hannah Arendt*, Thousand Oaks: Sage Publications.

Benjamin M. (2017) 'America Dropped 26,171 Bombs in 2016. What a Bloody End to Obama's Reign', *The Guardian*, 9 January, www.theguardian.com/commentisfree/2017/jan/09/america-dropped-26171-bombs-2016-obama-legacy.

Benjamin W. (2007) 'Critique of Violence', in Demetz P. (ed.), *Reflections – Essays, Aphorisms, Autobiographical Writing*, New York: Schocken Books: 277–300.

Bennett, J. (2010), *Vibrant Matter: An Ecology of Things*, Durham, NC: Duke University Press.

Bergman-Rosamond A. and Phythian M. (2011) *War, Ethics and Justice: New Perspectives on a Post-9/11 World*, Abingdon: Routledge.

Berkowitz R. (2010) 'Human Being in an Inhuman Age', Conference theme and address, Bard College, New York.

Berkowitz R. (2012) 'Human Being in an Inhuman Age', *HA: The Journal of the Hannah Arendt Center for Politics and Humanities* 1(1): 112–22.

Bernstein R. (2006) 'Rethinking the Social and the Political', in Williams, G. (ed.), *Hannah Arendt – Critical Assessments of Leading Political Philosophers*, Vol. III, Abingdon: Routledge.

Bernstein R.J. (1992) *The New Constellation: The Ethical-Political Horizons of Modernity/Postmodernity*, Cambridge, MA: MIT Press.

Berry D.M. and Pawlik J. (2005) 'What Is Code? A Conversation with Deleuze, Guattari and Code', *Kritikos: An International and Interdisciplinary Journal of Postmodern Cultural Sound, Text and Image* 2, http://intertheory.org/berry.htm.

Bertani M. and Fontana A. (2004) 'Situating the Lectures', in Foucault M., *Society Must Be Defended: Lectures at The College de France 1975–1976*, London: Penguin: 273–93.

Bigo D. (2006) 'Globalized (In)Security: the Field and the Ban-opticon', www.ces.fas.harvard.edu/conferences/muslims/Bigo.pdf.

Bigo D. (2011) 'Security: A Field Left Fallow', in Dillon M. and Neal A. (eds), *Foucault on Politics, Security and War*, Basingstoke: Palgrave Macmillan.

Birmingham P. (1994) 'Arendt/Foucault: Power and the Law', in Dallery A.B. and Watson S.H. (eds), *Transitions in Continental Philosophy*, Albany: State University of New York Press: 21–33.

Bland E. (2009) 'Robot Warriors Will Get a Guide to Ethics', *MSNBC*, 18 May, www.nbcnews.com/id/30810070/ns/technology_and_science-science/t/robot-warriors-will-get-guide-ethics.

Blencowe C. (2010) 'Foucault's and Arendt's "insider view" of Biopolitics: A Critique of Agamben', *History of the Human Sciences* 23(5): 113–30.

Bornstein J. (2015) 'U.S. Department of Defense Autonomy Roadmap: Autonomy Community of Interest', www.defenseinnovationmarketplace.mil/resources/AutonomyCOI_NDIA_Briefing20150319.pdf.

Bostrom N. (2005) 'Transhumanist Values', *Journal of Philosophical Research* 30 (Supplement): 3–14.

Bostrom N. (2014) *Superintelligence: Paths, Dangers, Strategies*, Oxford: Oxford University Press.

Bostrom N. (2015) 'Briefing on Existential Risk for the UN Interregional Crime and Justice Research Institute', New York: UN Interregional Crime and Justice Institute, http://webtv.un.org/watch/chemical-biological-radiologica l-and-nuclear-cbrn-national-action-plans-rising-to-the-challenges-of-internat ional-security-and-the-emergence-of-artificial-intelligence/4542739995001.

Bowcott O. (2017), ' "Specific" Terror Evidence Not Needed for RAF Drone Strikes', *The Guardian*, 11 January, www.theguardian.com/world/2017/jan/11/raf-dron e-strikes-terror-attorney-general.

Bowler P.J. (1989) *Evolution – The History of an Idea*, Berkeley: University of California Press.

Boyle M. (2015) 'The Legal and Ethical Implications of Drone Warfare', *The International Journal of Human Rights* 19(2): 105–26.

Braidotti R. (2011) *Nomadic Theory – The Portable Rosi Braidotti*, New York and Chichester: Columbia University Press.

Braidotti R. (2013) *The Posthuman*, Cambridge: Polity Press.

Braun K. (2007) 'Biopolitics and Temporality in Arendt and Foucault', *Time and Society* 16(5): 5–23.

Breen K. (2007) 'Violence and Power – A Critique of Hannah Arendt on the "Political"', *Philosophy and Social Criticsm* 33(3): 343–72.

Brennan J. (2010) 'Remarks by Assistant to the President for Homeland Security and Counterterrorism John Brennan at CSIS', The White House, www.whitehouse.gov/the-press-office/remarks-assistant-president-homelan d-security-and-counterterrorism-john-brennan-csi.

Brennan J. (2012a) 'The Efficacy and Ethics of U.S. Counterterrorism Strategy', Woodrow Wilson Center, www.wilsoncenter.org/event/the-efficacy-and-ethic s-us-counterterrorism-strategy.

Brennan J. (2012b) ' "This Week" Transcript: John Brennan, Economic Panel', *ABC News*, 29 April, http://abcnews.go.com/Politics/week-transcript-john-brennan/story?id=16228333andsinglePage=true#.T6QEOu1OXna.

Brennan J. (2013) 'Obama's Drone Master', *GQ*, www.gq.com/news-politics/big-issues/201306/john-brennan-cia-director-interview-drone-program?printable=true.

Brooks R. (2015) 'In Defense of Killer Robots', *Foreign Policy*, http://foreignpolicy.com/2015/05/18/in-defense-of-killer-robots/.

Brown C. (1992) *International Relations Theory – New Normative Approaches*, New York: Columbia University Press.

Brown C. (2002) *Sovereignty, Rights and Justice: International Political Theory Today*, Cambridge, MA: Polity.

Brown C. (2003) 'Self-Defense in an Imperfect World', *Ethics and International Affairs* 17(1): 2–8.

Bufacchi V. (2007) *Violence and Social Justice*, Basingstoke: Palgrave Macmillan.

Bufacchi V. (2009) *Violence, A Philosophical Anthology*, Basingstoke: Palgrave Macmillan.

Bull M. (2005) 'The Social and the Political', *South Atlantic Quarterly* 104(4): 675–92.

Bulley D. (2013) 'Foreign Policy as Ethics: Toward a Re-Evaluation of Values', *Foreign Policy Analysis* 10(2): 165–80.

Butler J. (2009a) 'Hannah Arendt, Ethics, and Responsibility', European Graduate School (EGS), video lecture, www.youtube.com/watch?v=vOwdsO6KkkI.

Butler J. (2009b) *Frames of War – When Is Life Grievable?*, London: Verso.

Byman D. (2013) 'Why Drones Work: The Case for Washington's Weapon of Choice', *Foreign Affairs* 92(4): 32–43.

Caldwell W. and Hagerott C.M.R. (2010) 'Curing Afghanistan', *Foreign Policy*, www.foreignpolicy.com/articles/2010/04/07/curing_afghanistan.

Calhoun C. and McGowan J. (eds) (1997) *Hannah Arendt and the Meaning of Politics*, Minneapolis: University of Minnesota Press.

Cameron D. (2011a) 'Full Transcript David Cameron Speech on the Fight-Back After The Riots', *New Statesman*, 15 August, www.newstatesman.com/politics/2011/08/society-fight-work-rights.

Cameron D. (2011b) 'PM Statement on Violence in England', *Speeches, GOV.UK.* www.gov.uk/government/speeches/pm-statement-on-violence-in-england.

Campbell D. (1998) 'Why Fight: Humanitarianism, Principles, and Post-structuralism', *Millennium Journal of International Studies* 27(3): 497–522.

Campbell D. (2001) 'Justice and International Order: The Case of Bosnia and Kosovo', in Coicaud J.-M. and Warner D. (eds), *Ethics and International Affairs – Extent and Limits*, New York: The United Nations University Press.

Campbell D. and Dillon M. (1993) *The Political Subject of Violence*, Manchester: Manchester University Press.

Campbell D. and Shapiro M.J. (1999) *Moral Spaces: Rethinking Ethics and World Politics*, Minneapolis: University of Minnesota Press.

Campbell T.C. (2011) *Improper Life: Technology and Biopolitics from Heidegger to Agamben*, Posthumanities, Minneapolis: University of Minnesota Press.

Campbell T.C. and Sitze A. (eds) (2013a) *Biopolitics: A Reader*, Durham, NC: Duke University Press.

Campbell T.C. and Sitze A. (2013b) 'Introduction. Biopolitics: An Encounter', in Campbell T.C. and Sitze A. (eds) *Biopolitics: A Reader*, Durham, NC: Duke University Press: 1–40.

Canovan M. (1995) *Hannah Arendt – A Reinterpretation of Her Political Thought*, Cambridge: Cambridge University Press.

Canovan M. (1998) 'Introduction', in Arendt H., *The Human Condition*, 2nd ed., Chicago: University of Chicago Press: vii–xx.

Carney J. (2013) 'Drone Strikes Are Legal, They Are Ethical, and They Are Wise', available from www.youtube.com/watch?v=Gg8jPheXtasandfeature=youtube_gdata_player.

Carpenter C. (2013) 'Beware the Killer Robots: Inside the Debate over Autonomous Weapons', *Foreign Affairs*, 3 July, www.foreignaffairs.com/articles/united-states/2013–07–03/beware-killer-robots.

Catalano Evers E., Fish L., Horowitz M.C. et al. (2017) 'Drone Proliferation: Policy Choices for the Trump Administration', *Papers for the President*, Washington, DC: Center for a New American Security.

Chadwick R. and Schroeder D. (eds) (2002) *Applied Ethics: Critical Concepts in Philosophy*, London: Routledge.

Chamayou G. (2015) *A Theory of the Drone*, New York: The New Press.

Chen S. (2017) 'China Unveils Its Answer to US Reaper Drone – How Does It Compare?' *South China Morning Post*, 17 July, www.scmp.com/news/china/diplomacy-defence/article/2103005/new-chinese-drone-oversea s-buyers-rival-us-reaper.

Cheney-Lippold J. (2011) 'A New Algorithmic Identity – Soft Biopolitics and the Modulation of Control', *Theory, Culture and Society* 28(6): 164–81.

Cherlin R. (2013) 'Obama's Drone Master', *GQ*, 17 June, www.gq.com/news-politics/big-issues/201306/john-brennan-cia-director-interview-drone-program?printable=true.

Coady C.A.J. (2009) 'The Idea of Violence', in Bufacchi V. (ed.), *Violence, A Philosophical Anthology*, Basingstoke: Palgrave Macmillan.

Coicaud J.M. and Warner D. (eds) (2001) *Ethics and International Affairs – Extent and Limits*, New York: The United Nations University Press.

Coker C. (2008) *Ethics and War in the Twenty-First Century*, London: Routledge.

Coker C. (2013) *Warrior Geeks: How the 21st Century Technology Is Changing the Way We Fight and Think About War*, New York: Columbia University Press.

Coker C. (2015) *Future War*, Cambridge: Polity.

Cole D. (2014) 'We Kill People Based on Metadata', *New York Review of Books*, 10 May, www.nybooks.com/daily/2014/05/10/we-kill-people-based-metadata.

Conteh-Morgan E. (2003) *Collective Political Violence: An Introduction to the Theories and Cases of Violent Conflicts*, New York: Routledge.

Cook M. (2011) Military Ethics as Professional Ethics: The Limits of the Philosophical Approach, The Oxford Institute for Ethics, Law and Armed Conflict, https://podcasts.ox.ac.uk/military-ethics-professional-ethics-limits-philosophical-approach.

Coombs J. and Winkler E. (1993) *Applied Ethics – A Reader*, Cambridge, MA: Blackwell.

Cooter R. and Stein C. (2010) 'Cracking Biopower', *History of the Human Sciences* 23(2): 109–28.

Crawford N.C. (2013) 'Bugsplat: US Standing Rules of Engagement, International Humanitarian Law, Military Necessity and Noncombatant Immunity', in Lang A.F., O'Driscoll C. and Williams J. (eds), *Just War: Authority, Tradition and Practice*, Washington, DC: Georgetown University Press.

Critchley S. (2009) *Ethics-Politics-Subjectivity*, London: Verso.

Critchley S. (2011) 'The Cycle of Revenge', *New York Times*, Opinionator, 9 August, http://opinionator.blogs.nytimes.com/2011/09/08/the-cycle-of-revenge.

Cudworth E. and Hobden S. (2011) *Posthuman International Relations: Complexity, Ecologism and Global Politics*, London: Zed Books.

Cummings M.L. (2017) 'Artificial Intelligence and the Future of Warfare', Research Paper, London: Chatham House, January.

Dauphinee E. (2009) 'Emanuel Levinas', in Edkins J. and Vaughan-Williams P.N. (eds), *Critical Theorists and International Relations*, Abingdon: Routledge.

Davis M. (2002) 'Thinking like an Engineer: The Place of a Code of Ethics in the Practice of a Profession', in Chadwick R. and Schroeder D. (eds), *Applied Ethics – Critical Concepts in Philosophy*, London: Routledge.

De Lazari-Radek K. and Singer P. (2012) 'The Objectivity of Ethics and the Unity of Practical Reason', *Ethics* 123(1): 9–31.

De Leonardis F. (2008) 'War as a Medicine: The Medical Metaphor in Contemporary Italian Political Language', *Social Semiotics* 18(1): 33–45.

Deleuze G. (2006) *Foucault*, London: Continuum.

D'Entreves M.P. (1994) *The Political Philosophy of Hannah Arendt*, London: Routledge.

D'Entreves M.P. (2006) 'Excerpt from "Hannah Arendt's Conception of Modernity"', in Williams G. (ed.), *Hannah Arendt – Critical Assessments of Leading Political Philosophers*, Vol. III, Abingdon: Routledge.

Derrida J. (1999) *Adieu to Emmanuel Levinas*, Stanford: Stanford University Press.

Derrida J. (2002) 'Force of Law', in Anidjar G. (ed.), *Acts of Religion*, London: Routledge: 228–98.

Derrida, J. (2004), 'Jacques Derrida, penseur de l'évènement', *L'Humanité*, 24 January, www.humanite.fr/node/299140.

Dewey J. (2009) 'Law, Violence and Force', in Bufacchi V. (ed.), *Violence, A Philosophical Anthology*, Basingstoke: Palgrave Macmillan.

De Wijze S. (2002) 'Toward a Political Ethic: Exploring the Boundaries of a Moral Politics', in Chadwick R. and Schroeder D. (eds), *Applied Ethics – Critical Concepts in Philosophy*, London: Routledge.

Dillon M. (1998) 'Criminalising Social and Political Violence Internationally', *Millennium Journal of International Studies* 27(3): 543–67.

Dillon M. (2007) 'Governing Terror: The State of Emergency of Biopolitical Emergence', *International Political Sociology* 1: 7–28.

Dillon M. (2011) 'Security, Race and War', in Dillon M. and Neal A. (eds), *Foucault on Politics, Security and War*, London: Palgrave Macmillan.

Dillon M. and Lobo-Guerro L. (2008) 'Biopolitics of Security in the 21st Century: An Introduction', *Review of International Studies* 34: 265–92.

Dillon M. and Neal A. (2011) *Foucault on Politics, Security and War*, London: Palgrave Macmillan.

Dillon M. and Reid J. (2009) *The Liberal Way of War – Killing to Make Life Live*, Abingdon: Routledge.

Diprose R. (2009) 'Toward an Ethico-Politics of the Posthuman: Foucault and Merleau-Ponty', *Parrhesia* 8: 7–19.

Doel M.A. (1999) *Poststructuralist Geographies: The Diabolical Art of Spatial Science*, Edinburgh: Edinburgh University Press.

Dolan F. (2005) 'The Paradoxical Liberty of Bio-Power: Hannah Arendt and Michel Foucault on Modern Politics', *Philosophy and Social Criticism* 31(3): 369–80.

Doyle M. (1983) 'Kant, Liberal Legacies, and Foreign Affairs', *Philosophy and Public Affairs* 12(3): 205–35.

Duarte A. (2006) 'Biopolitics and the Dissemination of Violence', in Williams G. (ed.), *Hannah Arendt – Critical Assessments of Leading Political Philosophers*, Vol. III, Abingdon: Routledge.

Dupuy J.P. (2007) 'Some Pitfalls in the Philosophical Foundations of Nanoethics', *Journal of Medicine and Philosophy* 32: 237–61.

Edwards C. (1999) 'Cutting off the King's Head: The "Social" in Hannah Arendt and Michel Foucault', *Studies in Social and Political Thought* 1(1): 3–20.

Ellul J. (1978) 'Symbolic Function, Technology and Society', *Journal of Social Biological Structures* 1: 207–18.

Endsley, M. (2015) 'Autonomous Horizons: Systems Autonomy in the Air Force – A Path to the Future. Volume I: Human-Autonomy-Teaming', United States Air Force Office of the Chief Scientist, www.af.mil/Portals/1/documents/SECAF/AutonomousHorizons.pdf?timestamp=1435068339702.

Enemark C. (2014) *Armed Drones and the Ethics of War: Military Virtue in a Post-Heroic Age*, Abingdon: Routledge.

Enns D. (2007) 'Political Life Before Identity', *Theory and Event* 10(1), http://muse-jhu-edu/article/213862.

Epstein R. (2016) 'The Empty Brain', *Aeon*, https://aeon.co/essays/your-brain-does-not-process-information-and-it-is-not-a-computer.

Esposito R. (2008) *Bios – Biopolitics and Philosophy*, Minneapolis: University of Minnesota Press.

European RPAS Steering Group (2013) *Roadmap for the Integration of Civil Remotely-Piloted Aircraft Systems into the European Aviation System*, Brussels: European Commission.

Evans B. (2013) *Liberal Terror*, Cambridge: Polity Press.

Fagan, M. (2013) *Ethics and Politics After Poststructuralism*, Edinburgh: Edinburgh University Press.

Fassin D. (2009) 'Another Politics of Life Is Possible', *Theory, Culture and Society* 26(5): 44–60.

Fenichel-Pitkin H. (1998) *The Attack of the Blob: Hannah Arendt's Concept of the Social*, Chicago: University of Chicago Press.

Finlay C. (2009) 'Hannah Arendt's Critique of Violence', *Thesis Eleven* 97: 26–45.

Finn J. (2014) 'Review of "Liquid Surveillance: A Conversation"', *Canadian Journal of Communication* 39(2): 1–2.

Fletcher G. (2000) 'The Nature and Function of Criminal Law', *California Law Review* 88(3): 687–703.

Foucault M. (1991) *Discipline and Punish – The Birth of the Prison*, London: Penguin.

Foucault M. (1998) *The Will to Knowledge: The History of Sexuality: 1*, London: Penguin Books.

Foucault M. (2000) 'On the Genealogy of Ethics: An Overview of Work in Progress', in Rabinow P. (ed.), *Ethics – Essential Works of Foucault* 1954–1984, London: Penguin.

Foucault M. (2002a) 'Governmentality', in Faubion J.D. (ed.), *Power – Essential Works of Foucault* 1954–1984, London: Penguin.

Foucault M. (2002b) 'The Subject and Power', in Faubion J.D. (ed.), *Power – Essential Works of Foucault* 1954–1984, London: Penguin.

Foucault M. (2004) *Society Must Be Defended*, London: Penguin.

Foucault M. (2007) *Security, Territory, Population*, Basingstoke: Palgrave Macmillan.

Frazer E. and Hutchings K. (2008) 'On Politics and Violence: Arendt Contra Fanon', *Contemporary Political Theory* 7: 90–108.

Frey R.G. and Heath Wellman C. (2005) *A Companion to Applied Ethics*, Malden, MA: Blackwell.

Frost M. (1996) *Ethics in International Relations: A Constitutive Theory* Cambridge: Cambridge University Press.

Frost M. (2009) *Global Ethics: Anarchy, Freedom and International Relations*, Abingdon: Routledge.

Fuller C.J. (2014) 'The Eagle Comes Home to Roost: The Historical Origins of the CIA's Lethal Drone Program', *Intelligence and National Security* 30(6): 769–92.

Gallagher S. (2016) 'U.S. DoD Continues Quest to Make "Iron Man" Exosuit for Special Ops', *Ars Technica UK*, https://arstechnica.co.uk/information-technology/2016/06/dod-continues-quest-to-make-iron-man-exosuit-for-special-ops.

Galliott J. and Lotz M. (eds) (2015) *Super Soldiers: Ethical, Legal and Social Implications*, Farnham: Ashgate.

Galtung J. (2009) 'Violence, Peace and Peace Research', in Bufacchi V. (ed.), *Violence, A Philosophical Anthology*, Basingstoke: Palgrave Macmillan.

Gamble A. (2010) 'Ethics and Politics', in Bell D. (ed.), *Ethics and World Politics*, New York: Oxford University Press.

Gaskarth J. (ed.) (2015) *Rising Powers, Global Governance and Global Ethics*, Abingdon: Routledge.

Gert B. (1984) 'Moral Theory and Applied Ethics', *The Monist* 67(4): 532–48.

Gertler J. (2012) 'U.S. Unmanned Aerial Systems', CRS Report for Congress, Congressional Research Service, www.crs.gov.

Gettinger D. (2016) 'Drone Spending in the Fiscal Year 2017 Defense Budget', New York: Centre for the Study of the Drone, http://dronecenter.bard.edu/files/2016/02/DroneSpendingFy17_CSD_1–1.pdf.

Gibbs S. (2017) 'Elon Musk: Regulate AI to Combat "Existential Threat" Before It's Too Late', *The Guardian*, 17 July. www.theguardian.com/technology/2017/jul/17/elon-musk-regulation-ai-combat-existential-threat-tesla-spacex-ceo.

Giddens A. (1987) *The Nation-State and Violence*, Berkeley and Los Angeles: University of California Press.

Giddens A. (2002) 'Political Theory and the Problem of Violence', in The Belgarde Circle (ed.), *The Politics of Human Rights*, London: Verso.

Giddens A. (2003) *Modernity and Self-Identity: Self and Society in the Late Modern Age*, Cambridge: Polity.

Golovina M. and Wroughton L. (2013) 'Kerry Hopes Pakistan Drone Strikes to End "Very Soon"', Reuters, 1 August, http://uk.reuters.com/article/2013/08/01/uk-usa-pakistan-kerry-idUKBRE97015820130801.

Goodhand J. and Hulme D. (1999) 'From Wars to Complex Political Emergencies: Understanding Conflict and Peace-Building in the New World Disorder', *Third World Quarterly* 20(1): 13–26.

Gordon N. (2002) 'On Visibility and Power: An Arendtian Corrective of Foucault', *Human Studies* 25: 125–45.

Grayson K. (2012) 'The Ambivalence of Assassination: Biopolitics, Culture and Political Violence', *Security Dialogue* 43(1): 25–41.

Gregory D. (2008) 'The Rush to the Intimate: Counterinsurgency and Culture in the Late Modern War', *Radical Philosophy* 150(8): 8–23.

Gregory D. (2011a) 'The Everywhere War', *The Geographical Journal* 177(3): 238–50.

Gregory D. (2011b) 'From a View to Kill: Drones and Late Modern War', *Theory, Culture and Society* 28(7–8): 188–215.

Grossman D. (1996) *On Killing: The Psychological Cost of Learning to Kill in War and Society*, London: Little, Brown and Co.

Halberstam J. and Livingston I. (1995) *Posthuman Bodies*, Bloomington: Indiana University Press.

Hanssen B. (2000) *Critique of Violence: Between Post-Structuralism and Critical Theory*, London: Routledge.

Haraway D. (1997) *Modest_Witness@Second_Millennium.FemaleMan_Meets_OncoMouse: Feminism and Technoscience*, New York: Routledge.

Hardt M. and Negri A. (2001) *Empire*. Cambridge, MA: Harvard University Press.

Hardt M. and Negri A. (2006) *Multitude*, London: Penguin Books.

Hassner P. (2001) 'Violence and Ethics: Beyond the Reason of State Paradigm', in Coicaud J.-M and Warner D. (eds), *Ethics and International Affairs – Extent and Limits*, New York: The United Nations University Press.

Hayden P. (2009) *Political Evil in a Global Age*, Oxford: Routledge.

Heisenberg W. (1958) *The Physicist's Conception of Nature*, London: Hutchinson and Co.

Hobsbawm E. (2007) *Globalisation, Democracy and Terrorism*, London: Little, Brown and Co.

Høivik T. and Galtung J. (1971) 'Structural and Direct Violence', *Journal of Peace Research* 8(1): 73–6.

Holloway C. and Gruber K. (2003) 'Peak Soldier Performance', *Perspective, Science and Technology at Strategy Analysis, Incorporated*, 4th Quarter, www.sainc.com/TechnicalReports/download/4QTR03.pdf.

Honig B. (1995) 'Towards an Agonistic Feminism', in Honig B. (ed.), *Feminist Interpretations of Hannah Arendt*, University Park: Pennsylvania State University Press: 135–66.

Honig B. (2006) 'Toward an Agonistic Feminism: Hannah Arendt and the Politics of Identity', in Williams G. (ed.), *Hannah Arendt – Critical Assessments of Leading Political Philosophers*, Vol. III, Abingdon: Routledge.

Howell A. (2014) 'The Global Politics of Medicine: Beyond Global Health, Against Securitisation Theory', *Review of International Studies* 40: 961–87.

Human Rights Watch (2012) *Losing Humanity – The Case against Killer Robots*, Cambridge, MA: International Human Rights Clinic at Harvard Law School.

Hutchings K. (2010) *Global Ethics – An Introduction*, Cambridge: Polity Press.

Illich I. (1975) 'Clinical Damage, Medical Monopoly, the Expropriation of Health: Three Dimensions of Iatrogenic Tort', *Journal of Medical Ethics* 1(2): 78–80.

Illich, I. (1976). *Medical Nemesis: The Expropriation of Health*, New York: Random House.

Isaac J.C. (1992) *Arendt, Camus and Modern Rebellion*, New Haven: Yale University Press.

Isaac J.C. (1996) 'A New Guarantee on Earth: Hannah Arendt on Human Dignity and the Politics of Human Rights', *The American Political Science Review* 90(1): 61–73.

Jabri V. (1998) 'Restyling the Subject of Responsibility in International Relations', *Millennium Journal of International Studies* 27(3): 591–612.

Jabri V. (2006) 'War, Security and the Liberal State', *Security Dialogue* 37(1): 47–64.

Jabri V. (2007) *War and the Transformation of Global Politics*, Basingstoke: Palgrave Macmillan.

Jacobsen M.H. and Poder P. (2008) *The Sociology of Zygmunt Bauman: Challenges and Critique*, Farnham: Ashgate Publishing.

Jay M. (2006) 'The Political Existentialism of Hannah Arendt', in Williams G. (ed.), *Hannah Arendt – Critical Assessments of Leading Political Philosophers*, Vol. III, Abingdon: Routledge.

Johansson L. (2011) 'Is It Morally Right to Use Unmanned Aerial Vehicles (UAVs) in War?', *Philosophy and Technology* 24: 279–91.

Joyce R. (2006) *The Evolution of Morality*, Cambridge, MA: MIT Press.

Kaag J. and Kreps S. (2012a) 'The Use of Unmanned Aerial Vehicles in Contemporary Conflict: A Legal and Ethical Analysis', *Polity* 44: 260–85.

Kaag J. and Kreps S. (2012b) 'Drones Bring an End to War's Easy Morality', *The Chronicle of Higher Education*, www.chronicle.com/article/Opinion-Drones-End-Wars-Easy/134236.

Kateb G. (1984) *Hannah Arendt – Politics, Conscience, Evil*, Totowa, NJ: Rowman and Allenheld.

Keane J. (2004) *Violence and Democracy*, Cambridge: Cambridge University Press.

Kennedy C. and Rogers J. (2015) 'Virtuous Drones?', *International Journal of Human Rights* 19(2): 211–27.

King D. (n.d.) 'The Luddites and Bio Politics Now', The Luddites at 200: For Action, Against Technology 'Hurtful to Commonality', www.luddites200.org.uk/biopolitics.html#sdendnote1anc/.

Koch B. and Schöring N. (2015) 'The Dangers of Lethal Autonomous Weapons Systems', *Justitia et Pax Europa*, 22 November, www.justitiaetpax.dk/nyheder/2015/the-dangers-of-lethal-autonomous-weapons-systems.

Koch C. (2012) 'How Physics and Neuroscience Dictate Your "Free" Will: Scientific American', *Scientific American*, www.scientificamerican.com/article.cfm?id=finding-free-will.

Kochi T. (2009) *The Other's War – Recognition and the Violence of Ethics*, Oxford: Birkbeck Law Press.

Kohn J. (1990) 'Thinking/Acting', *Social Research: An International Quarterly* 57(1): 105–34.

Kohn J. (2003) 'Introduction', in Kohn J. (ed.), *Responsibility and Judgment*, New York: Schocken: vii–xxix.

Kohn J. and May L. (eds) (1997) *Hannah Arendt – Twenty Years Later*, Cambridge, MA: MIT Press.

Kreide R. (2009) 'Preventing Military Humanitarian Intervention? John Rawls and Juergen Habermas on a Just Global Order', *German Law Journal – Special Issue: The Kantian Project of International Law* 10(1): 363–81.

Kroker A. (2014), *Exits to the Posthuman Future*, Cambridge: Polity.

Kurzweil R. (2006) *The Singularity Is Near: When Humans Transcend Biology*, London: Gerald Duckworth.

Kurzweil R. (2016) 'Superintelligence and Singularity', in Schneider S. (ed.), *Science Fiction and Philosophy from Time Travel to Superintelligence*, Chichester: John Wiley and Sons.

Kymlicka W. and Sullivan W.M. (eds) (2007) *The Globalization of Ethics*, New York: Cambridge University Press.

LaFollette H. (2003) *The Oxford Handbook of Practical Ethics*, Oxford: Oxford University Press.

Lakoff G. (1991) 'Metaphor and War: The Metaphor System Used to Justify the War in the Gulf', *Viet Nam Generation Journal and Newsletter* 3(3), www2.iath.virginia.edu/sixties/HTML_docs/Texts/Scholarly/Lakoff_Gulf_Metaphor_1.html.

Latour B. (2005) *Reassembling the Social: An Introduction to Actor-Network Theory*, Oxford: Oxford University Press.

Lazar S. (2011) 'Morality and Law in War', in Marmor A. (ed.), *The Routledge Companion to Philosophy of Law*, New York: Routledge: 364–79.

Lazar S. (2013) 'Legitimate Authority and the Morality of War', Presentation: London School of Economics, handout.

Lazzarato M. (2002) 'From Biopower to Biopolitics', *The Warwick Journal of Philosophy* 13: 1–10.

Leander A. (2013) 'Technological Agency in the Co-Constitution of Legal Expertise and the US Drone Program', *Leiden Journal of International Law* 26(4): 811–31.

Lee P. (2013) 'Debate: Is the Use of Unmanned Military Drones Ethical or Criminal?', *The Guardian*, 2 December, www.theguardian.com/commentisfree/video/2013/dec/02/unmanned-military-drones-battle-ethical-video-debate.

Lin P. (2011) 'Drone-Ethics Briefing: What a Leading Robot Expert Told the CIA', *The Atlantic*, www.theatlantic.com/technology/archive/2011/12/drone-ethics-briefing-what-a-leading-robot-expert-told-the-cia/250060/.

Litke R.F. (2009) 'Violence and Power', in Bufacchi V. (ed.), *Violence, A Philosophical Anthology*, Basingstoke: Palgrave Macmillan.

Lorenz A., Mittelstaedt J. von and Schmitz G.P. (2011) 'Are Drones Creating a New Global Arms Race?', *Spiegel Online*, 21 October, www.spiegel.de/international/world/0,1518,792590,00.html.

Ludz U. (2007) 'Arendt's Observations and Thoughts on Ethical Questions', *Social Research* 74(3): 797–810.

Lupton J.R. (2012) 'Arendt in Italy: Or, the Taming of the Shrew', *Law, Culture and the Humanities* 8(1): 68–83.

MacIntyre A. (1981) *After Virtue – A Study in Moral Theory*, London: Gerald Duckworth and Co. Ltd.

MacIntyre A. (1984) 'Does Applied Ethics Rest on a Mistake?', *The Monist* 67(4): 498–513.

Manchev B. (2009) 'Terror and the Crisis of the Political', in Das S.K. (ed.). *Terror, Terrorism, States and Societies*, Delhi: Women Unlimited.

Marcus G. (2012) 'Moral Machines', *The New Yorker*, 24 November, www.newyorker.com/online/blogs/newsdesk/2012/11/google-driverless-car-morality.html.

Marcuse H. (1982) 'Some Social Implications of Modern Technology', in Arato A. and Gebhardt E. (eds), *The Essential Frankfurt School Reader*, New York: The Continuum International Publishing Group: 138–62.

Markoff J. (2010) 'Robots, the Military's Newest Forces', *The New York Times*, 27 November, www.nytimes.com/2010/11/28/science/28robot.html.

Mbembe A. (2003) 'Necropolitics', *Public Culture* 15(1): 11–40.

McFalls L. (2007) A Matter of Life and Death: Iatrogenic Violence and the Formal Logic of International Intervention. Trent University, Peterborough, Ontario: Centre for the Critical Study of Global Power and Politics.

McMahan J. (2004) 'The Ethics of Killing in War', *Ethics* 114: 693–733.

McMahan J. (2008) 'The Morality of War and the Law of War', in Rodin D. and Shue H. (eds), *Just and Unjust Warriors: The Moral and Legal Status of Soldiers*, Oxford: Oxford University Press.

McMahan J. (2009) *Killing in War*, Oxford: Oxford University Press.

McMahan J. (2011) 'Duty, Obedience, Desert and Proportionality in War: A Response', *Ethics* 122(1): 135–67.

McMahan J. (2013) 'Where to Now for Just War Theory?', The Oxford Institute for Ethics, Law and Armed Conflict. Podcast.

McNeal G.S. (2011) The U.S. Practice of Collateral Damage Estimation and Mitigation, Pepperdine Working Paper, Pepperdine University, http://works.bepress.com/gregorymcneal/22/.

Meade E.M. (1997) 'The Commodification of Values', in May L. and Kohn J. (eds), *Hannah Arendt: Twenty Years Later*, Cambridge, MA: MIT Press.

Meffan J. and Worthington K.L. (2001) 'Ethics before Politics', in Davis T.F. and Vomack K. (eds), *Mapping the Ethical Turn: A Reader in Ethics, Culture, and Literary Theory*, Charlottesville: University Press of Virginia.

Mehlman M., Lin P. and Abney K. (2013) 'Enhanced Warfighters: Risk, Ethics, and Policy'. Case Legal Studies Research Paper No. 2013-2. http://dx.doi.org/10.2139/ssrn.2202982.

Miller G. (2012) 'White House Approves Broader Yemen Drone Campaign', *The Washington Post*, 25 April, www.washingtonpost.com/world/national-security/white-house-approves-broader-yemen-drone-campaign/2012/04/25/gIQA82U6hT_story.html?sub=AR.

Miller K. (2014), 'Total Surveillance, Big Data and Predictive Crime Technology: Privacy's Perfect Storm', *Journal of Technology of Law and Policy* 19: 105–46.

Mitchell A. (2014) 'Dispatch from Robot Wars: Or, What Is Posthuman Security', *The Disorder of Things*, https://thedisorderofthings.com/2014/07/24/dispatches-from-the-robot-wars-or-what-is-posthuman-security.

Mooney J. and Young J. (2005) 'Imagining Terrorism: Terrorism and Anti-Terrorism Terrorism, Two Ways of Doing Evil', *Social Justice* 32(1): 113–25.

Mouffe C. (2005) *The Return of the Political*, London: Verso.

Moyes R. and Roff H. (2016) 'Meaningful Human Control, Artificial Intelligence and Autonomous Weapons', in Article 36, *Geneva*,www.article36.org/wp-content/uploads/2016/04/MHC-AI-and-AWS-FINAL.pdf.

Mudd P. (2012) 'The Limits of Drone Warfare. 3rd August', www.thedailybeast.com/articles/2012/08/03/the-limits-of-drone-warfare.html.

Muehlhauser L. and Helm L. (2012) 'Intelligence Explosion and Machine Ethics', in Eden A., Soraker J., Moor J.H. et al. (eds), *Singularity Hypotheses: A Scientific and Philosophical Assessment*, Berlin: Springer.

Mueller J.M. (2002) 'Non-Violence in Education', UNESCO Publications, http://portal.unesco.org/education/en/file_download.php/fa99ea234f4accb0ad43040e1d60809cmuller_en.pdf.

Mumford L. (1934) *Technics and Civilisation*, New York: Harcourt, Brace and Co.

Murakami H. (2005) *Kafka on the Shore*, London: Random House Vintage.

Musk E. (2014) 'Interview with Walter Isaacson', *Vanity Fair, New Establishment Summit: The Age of Innovation*, 8 October, www.vanityfair.com/online/daily/2014/10/elon-musk-artificial-intelligence-fear.

Nabulsi K. (2006) 'Conceptions of Justice in War: From Grotius to Modern Times', in Sorabji R. and Rodin D. (eds), *The Ethics of War – Shared Problems in Different Traditions*, Aldershot: Ashgate.

Nardin T. (2008) 'International Ethics', in Reus-Smit C. and Snidal D. (eds), *Oxford Handbook of International Relations*, Oxford: Oxford University Press.

Nash W.L. (2002) 'The Laws of War: A Military View', *Ethics and International Affairs* 16(1): 14–17.

National Academies (2008) *Emerging Cognitive Neuroscience and Related Technologies*, Washington, DC: National Research Council Committee on Military and Intelligence Methodology for Emergent Neurophysiological and Cognitive/Neural Research.

New America Foundation (2012) 'The Drone War in Pakistan', *New America Foundation*, http://natsec.newamerica.net/drones/pakistan/analysis.

New America Foundation (2017) 'Who Has What: Countries with Armed Drones', www.newamerica.org/in-depth/world-of-drones/3-who-has-what-countries-armed-drones.

Nietzsche F. (1973) *Genealogie der Moral – Werke in drei Bänden.* 3, Munich: HanserVerlag.

Nolin P.C. (2012) *Unmanned Aerial Vehicles: Opportunities and Challenges for the Alliance*, NATO Parliamentary Assembly, NATO, www.nato-pa.int.

Obama B. (2013) Obama's speech on drone policy. *The New York Times*, 23 May, www.nytimes.com/2013/05/24/us/politics/transcript-of-obamas-speech-on-drone-policy.html?_r=0.

Obama B. (2015) 'Address to the Nation by the President', The White House, 6 December, https://obamawhitehouse.archives.gov/the-press-office/2015/12/06/address-nation-president.

O'Connell M.E. (2010) *Lawful Use of Combat Drones*, www.fas.org/irp/congress/2010_hr/042810oconnell.pdf.

O'Connell M. (2017) *To Be a Machine*. London: Granta Books.

Ofek H. (2010) 'The Tortured Logic of Obama's Drone War', *The New Atlantis* 27 (Spring): 35–44.

Oksala J. (2010) 'Violence and the Biopolitics of Modernity', *Foucault Studies* 10: 23–43.

Olson J. and Rashid M. (2013) 'Modern Drone Warfare: An Ethical Analysis', ASEE Southeast Section Conference, American Society for Engineering Education, presentation at Southeast Section Conference. 10–12 March, Cookeville, TN.

Open Roboethics Initiative (2015) *The Ethics and Governance of Lethal Autonomous Weapons Systems: An International Opinion Poll*, OpenRoboEthics, www.openroboethics.org/wp-content/uploads/2015/11/ORi_LAWS2015.pdf.

Orlie M. (1997) *Living Ethically, Acting Politically*, Ithaca: Cornell University Press.

Oudes C. and Zwijnenburg W. (2011) *Does Unmanned Make Unacceptable? Exploring the Debate on Using Drones and Robots in Warfare*, Utrecht: IKV Pax Christi.

Owens P. (2007) *Between Wars and Politics*, New York: Oxford University Press.

Owens P. (2009) 'Reclaiming "Bare Life"?: Against Agamben on Refugees', *International Relations* 23(4): 567–82.

Owens P. (2012) 'Human Security and the Rise of the Social', *Review of International Studies* 38(3): 547–67.

Papadopoulos D. (2010) 'Insurgent Posthumanism', *Ephemera – Theory and Politics in Organization* 10(2): 134–51.

Parekh S. (2008) *Hannah Arendt and the Challenge of Modernity*, New York: Routledge.

Parker T. (2012) 'Drones: The Known Knowns', *Human Rights Now*, Amnesty International, http://blog.amnestyusa.org/us/drones-the-known-knowns/.

Parsons D. (2013) 'Worldwide, Drones Are in High Demand', *National Defense*, www.nationaldefensemagazine.org/archive/2013/May/Pages/Worldwide, DronesAreinHighDemand.aspx.

Pateman C. (1986) 'Removing Obstacles to Democracy', paper presented to the International Studies Association meeting, Ottawa, Canada, October.

Patton P. (1998) 'Foucault's Subject of Power', in Moss J. (ed.), *The Later Foucault: Politics and Philosophy*, London: Sage.

Pugliese J. (2011) 'Prosthetics of Law and the Anomic Violence of Drones', *Griffith Law Review* 20(4): 931–61.

Raffoul F. (2008) 'Derrida and the Ethics of the Im-possible', *Research in Phenomenology* 38(2): 270–90.

Rapoport D.C. and Weinberg L. (eds) (2001) *The Democratic Experience and Political Violence*, London: Frank Cass.

Reid J. (2006) *The Biopolitics of the War on Terror: Life Struggles, Liberal Modernity, and the Defence of Logistical Societies*, Manchester: Manchester University Press.

Rengger N. (2002) 'On the Just War Tradition in the Twenty-First Century', *International Affairs* 78(2): 353–63.

Rengger N. (2013) *Just War and International Order: The Uncivil Condition in World Politics*, Cambridge: Cambridge University Press.

Ricoeur P. (2006) 'Power and Violence', in Williams G. (ed.), *Hannah Arendt – Critical Assessments of Leading Political Philosophers*, Vol. III, Abingdon: Routledge.

Rodd R. (1990) *Biology, Ethics and Animals*, Oxford: Clarendon.

Roden D. (2015) *Posthuman Life: Philosophy at the Edge of the Human*, Abingdon: Routledge.

Roderick I. (2010), 'Mil-Bot Fetishism: The Pataphysics of Military Robots', *Topia* 23–4: 286–303.

Rodin D. (2011) 'Justifying Harm', *Ethics* 122(1): 74–110.

Rodin D. and Sorabji R. (eds) (2006) *The Ethics of War – Shared Problems in Different Traditions*, Burlington: Ashgate.

Rose N. (2001) 'The Politics of Life Itself', *Theory, Culture and Society* 18(6): 1–30.

Rose N. (2003) *Powers of Freedom: Reframing Political Thought*, New York: Cambridge University Press.

Rose N. (2006) *Politics of Life Itself: Biomedicine, Power and Subjectivity in the Twenty-First Century*, Princeton: Princeton University Press.

Rosenberg M. and Haberman M. (2016) 'Michael Flynn, Anti-Islamist Ex-General Offered Security Post, Trump Aide Says', *The New York Times*, 17 November, www.nytimes.com/2016/11/18/us/politics/michael-flynn-national-security-adviser-donald-trump.html.

Roulier S. (1997) 'Excavating Foucaultian Identity', *Humanitas* X(1), www.nhinet.org/roulier.htm.

Royal Society (2012) *Neuroscience, Conflict and Security*, Brain Waves Module 3, London: Royal Society.

Sainato M. (2015) 'Stephen Hawking, Elon Musk, and Bill Gates Warn About Artificial Intelligence', *The Observer*, 19 August, http://observer.com/2015/08/stephe n-hawking-elon-musk-and-bill-gates-warn-about-artificial-intelligence/>.

Salamon J. (2015) *Solidarity Beyond Borders: Ethics in a Globalising World*, London: Bloomsbury Publishing.

Salmi J. (2009) 'The Different Categories of Violence', in Bufacchi V. (ed.), *Violence, A Philosophical Anthology*, Basingstoke: Palgrave Macmillan.

Sanders A. and Wolfgang M. (2014) 'The Rise of Robotics', *bcg.perspectives*, www.bcgperspectives.com/content/articles/business_unit_strategy_innovation_ rise_of_robotics.

Sanger D. (2011) 'As Drones Evolve, More Countries Want Their Own', National Security, National Public Radio (NPR), www.npr.org/2011/09/26/140812779/ as-drones-evolve-more-countries-want-their-own.

Sauer F. (2014) 'Autonomous Weapons Systems: Humanising of Dehumanising Warfare?', *Global Governance Spotlight* 4 (Bonn: Stiftung Entwicklung und Frieden).

Schaap A. (2007) 'Political Theory and the Agony of Politics', *Political Studies Review* 5(1): 56–74.

Schaap A. (2010) 'The Politics of Need', in Celemajer D. and Schaap A. (eds), *Power, Judgement and Political Evil: In Conversation with Hannah Arendt*, Farnham: Ashgate: 157–69.

Schwarz E. (2016), 'Prescription Drones: On the Techno-Biopolitical Regimes of Contemporary "Ethical Killing"', *Security Dialogue* 47(1): 59–75.

Shachtman N. (2007) 'Supercharging Soldier's Cells', *Wired*, www.wired.com/ 2007/03/supercharging_s/.

Shachtman N. (2012) 'Darpa's Magic Plan: "Battlefield Illusions" to Mess with Enemy Minds', Danger Room, *Wired*, www.wired.com/dangerroom/2012/02/ darpa-magic.

Sharkey N. (2010) 'Saying "No!" to Lethal Autonomous Targeting', *Journal of Military Ethics* 9(4): 369–83.

Sharkey N. (2012) 'America's Mindless Killer Robots Must Be Stopped', *The Guardian*, 3 December, www.guardian.co.uk/commentisfree/2012/dec/03/ mindless-killer-robots.

Shaw I. (2013) 'Predator Empire: The Geopolitics of US Drone Warfare', *Geopolitics* 18(3): 536–59.

Shaw I. (2016) *Predator Empire: Drone Warfare and Full Spectrum Dominance*, Minneapolis: University of Minnesota Press.

Shinko R.E. (2008) 'Agonistic Peace: A Postmodern Reading', *Millennium Journal of International Studies* 36(3): 473–91.

Shue H. (1995) 'Ethics, the Environment and the Changing International Order', *International Affairs* 71(3): 453–61.

Simon J.D. (2001) *The Terrorist Trap: America's Experience with Terrorism*, Bloomington: Indiana University Press.

Singer P.W. (2009) 'Robots at War: The New Battlefield', *The Wilson Quarterly – Surveying the World of Ideas*, www.cc.gatech.edu/classes/AY2009/cs4001c_ spring/documents/WilsonQuarterly-RobotsAtWar.pdf.

Singer P.W. (2010a) 'The Ethics of Killer Applications: Why Is It So Hard to Talk About Morality When It Comes to New Military Technology?', *Journal of Military Ethics* 9(4): 299–312.

Singer P.W. (2010b) *Wired for War: The Robotics Revolution and Conflict in the 21s Century*, London: Penguin.

Singer P.W. (2011) *The Morality of Robo-Wars: PW Singer*, 3 December, The Philosopher's Zone, Radio National.

Sloterdijk P. (2009) 'Rules for the Human Zoo: A Response to the Letter on Humanism', *Environment and Planning* 27: 12–28.

Small M. and Singer D.J. (1976), 'The War Proneness of Democratic Regimes, 1816–1965', *Jerusalem Journal of International Relations* 1: 50–69.

Spiro D.E. (1994), 'Give Democratic Peace a Chance? The Insignificance of the Liberal Peace', *International Security* 19(2): 50–86.

Statman D. (2015) 'Drones and Robots: On the Changing Practice of Warfare', in Lazar S. and Frowe H. (eds), *The Oxford Handbook of Ethics and War*, Oxford: Oxford University Press.

Stenger V. (2012) 'Free Will Is an Illusion', *Huffington Post*, 1 August, www.huffingtonpost.com/victor-stenger/free-will-is-an-illusion_b_1562533.html.

Stepanova E. (2009), 'One-Sided Violence Against Civilians in Armed Conflicts', in *SIPRI Yearbook 2009*, Oxford: Oxford University Press.

Stock G. (2002) *Redesigning Humans: Our Inevitable Genetic Future*, Boston: Houghton Mifflin.

Strawser B. (2010) 'Moral Predators: The Duty to Employ Uninhabited Aerial Vehicles', *Journal of Military Ethics* 9(4): 342–68.

Strawser B.J. (2013), 'Introduction', in Strawser B. (ed.), *Killing by Remote Control: The Ethics of an Unmanned Military*, New York: Oxford University Press: 3–24.

Sullins J.P. (2010) 'Robo Warfare: Can Robots Be More Ethical than Humans on the Battlefield?', *Ethics of Information Technology* 12: 263–75.

Swiffen A. (2011) *Law, Ethics and the Biopolitical*, Abingdon: Routledge.

Swift S. (2008) *Hannah Arendt*, Abingdon: Routledge.

Tamborino J. (1999) 'Locating the Body: Corporeality and Politics in Hannah Arendt', *The Journal of Political Philosophy* 7(2): 172–90.

Tancredi L. (2005) *Hardwired Behavior: What Neuroscience Reveals About Morality*, New York: Cambridge University Press.

Taylor D. (2007) 'Arendt, Foucault, and Feminist Politics: A Critical Appraisal', in Orr D. (ed.), *Feminist Politics – Identity, Difference, and Agency*, Lanham, MD: Rowman and Littlefield Publishers: 243–65.

Thacker E. (2003) 'Data Made Flesh: Biotechnology and the Discourse of the Posthuman', *Cultural Critique* 53(Winter): 72–97.

Till C. (2013) 'Architects of Time: Labouring on Digital Futures', *Thesis Eleven* 118(1): 33–47.

Tilly C. (2003) *The Politics of Collective Violence*, New York: Cambridge University Press.

Turda M. (2007) 'The Nation as Object: Race, Blood, and Biopolitics in Interwar Romania', *Slavic Review* 66(3): 413–41.

Turda M. (2010) *Modernism and Eugenics*, Basingstoke and New York: Palgrave Macmillan.

US Department of Defense (2012) U.S. Department of Defense Directive 3000.09, www.esd.whs.mil/Portals/54/Documents/DD/issuances/dodd/300009p.pdf

US Department of Defense (2017a) *Defense Budget Overview: United States Department of Defense Fiscal Year 2018 Budget Request*, Washington, DC: US Department of Defense.

US Department of Defense (2017b) *Unmanned Aircraft Systems (UAS): DoD Purpose and Operational Use*. US Department of Defense Search, www.defense.gov/UAS/.

Vatter M. (2006) 'Natality and Biopolitics in Hannah Arendt', *Revista de Ciencia Politica* 26(2): 137–59.

Vatter M. (2009) 'From Surplus Value to Surplus Life', *Theory and Event* 12(2), http://muse.jhu.edu/article/269986.

Villa D.R. (1999) *Politics, Philosophy, Terror*, Princeton: Princeton University Press.

Vincenzo B.N.M. (1992) 'Darwin on Man in the Origin of Species: Further Factors Considered', *Journal of the History of Biology* 25(1): 137–47.

Virilio P. (1986) *Speed and Politics: An Essay on Dromology*, New York: Columbia University.

Wagner M. (2012) 'Law, Ethics and the Biopolitical', http://robots.law.miami.edu/wp-content/uploads/2012/01/Wagner_Dehumanization_of_international_humanitarian_law.pdf.

Wald P. (2007) *Contagious: Cultures, Carriers, and the Outbreak Narrative*, Durham, NC: Duke University Press.

Walzer M. (2016) 'Just and Unjust Targeted Killing and Drone Warfare', *Daedalus: The Journal of the American Academy of Arts and Science* 145(5): 12–24.

Wan W. and Finn P. (2011) 'Global Race on to Match U.S. Drone Capabilities', *The Washington Post*, 8 July, www.washingtonpost.com/world/national-security/global-race-on-to-match-us-drone-capabilities/2011/06/30/gHQACWdmxH_story.html.

Weber J. (2016) 'Keep Adding: On Kill Lists, Drone Warfare and the Politics of Databases', *Environment and Planning D: Society and Space* 34(1): 107–25.

Weir L. (1996) 'Recent Developments in the Government of Pregnancy', *Economy and Society* 25(3): 373–92.

Weizman E. (2009) 'Thanato-Tactics', in Ophir A., Givoni M. and Hanafi S. (eds), *The Power of Inclusive Exclusion: Anatomy of Israeli Rule in the Occupied Territories*, New York: Zone Books.

Westphal C. (2016) 'Treating Terrorism like Cancer', *Boston Globe*, 2 April, www.bostonglobe.com/opinion/2016/04/01/treating-terrorism-like-cancer/6mSHClNC9U1fdvZOAj90PM/story.html.

Whetham D. (2013) 'Killer Drones: The Moral Ups and Downs', *RUSI Journal* 158(3): 22–32.

White House (2002) *The National Security Strategy of the United States of America*, Washington, DC: White House.

Widdows H. (2011) *Global Ethics*, Abingdon: Routledge.

Williams A. (2011) 'Enabling Persistent Presence? Performing the Embodied Geopolitics of the Unmanned Aerial Vehicle assemblage', *Political Geography* 30: 381–90.

Williams G. (ed.) (2006) *Hannah Arendt – Critical Assessments of Leading Political Philosophers*, Vol. III, New York: Routledge.

Williams J. (2015) 'Distant Intimacy: Space, Drones and Just War', *Ethics and International Affairs* 29(1): 93–110.

Willmott C. (2016) *Biological Determinism, Free Will and Moral Responsibility*, Springer Briefs in Ethics, London: Springer International Publishing.

Wilson E.O. (1998) 'The Biological Basis of Morality', The Atlantic Monthly – Digital Edition, www.theatlantic.com/past/docs/issues/98apr/biomoral.htm.

Wilson S. and Cohen J. (2012) 'Poll Finds Broad Support for Obama's Counterterrorism Policies', The Washington Post, 9 February, www.washingtonpost.com/politics/poll-finds-broad-support-for-obamas-counterterrorism-policies/2012/02/07/gIQAFrSEyQ_story.html.

Wittes B. and Byman D. (2013) 'How Obama Decides Your Fate if He Thinks You're a Terrorist', www.theatlantic.com/international/archive/2013/01/how-obama-decides-your-fate-if-he-thinks-youre-a-terrorist/266419.

Wolin S. (2009) 'Violence and the Western Political Tradition', in Bufacchi V. (ed.), *Violence, A Philosophical Anthology*, Basingstoke: Palgrave Macmillan.

Woods C. and Yusufzai M. (2013) 'Get the Data: The Return of Double-Tap Drone Strikes', The Bureau of Investigative Journalism, www.thebureauinvestigates.com/2013/08/01/get-the-data-the-return-of-double-tap-drone-strikes.

Worcester K., Bermanzohn S.A. and Ungar M. (2002) *Violence and Politics: Globalization's Paradox*, New York: Taylor and Francis Group.

Youde J. (2009) *Biopolitical Surveillance and Public Health in International Politics*, Basingstoke: Palgrave Macmillan.

Young-Bruehl E. (2006) *Why Arendt Matters*, New Haven: Yale University Press.

Zacharias G. (2015) 'US Armed Services Committee Hearing on "Advancing the Science and Acceptance of Autonomy for Future Defense Systems"', http://armedservices.house.gov/index.cfm/2015/11/advancing-the-science-and-acceptance-of-autonomy-for-future-defense-systems.

Zehfuss M. (2011) 'Targeting: Precision and the Production of Ethics', *European Journal of International Relations* 17(3): 543–66.

Zenko M. (2017) 'The (Not-So) Peaceful Transition of Power: Trump's Drone Strikes Outpace Obama', Council of Foreign Relations, 2 March, www.cfr.org/blog/not-so-peaceful-transition-power-trumps-drone-strikes-outpace-obama.

Index

Lightning Source UK Ltd.
Milton Keynes UK
UKHW022116260320
360939UK00007B/339

9 781526 114846